ADVANCED COMMUNICATIONS AND MULTIMEDIA SECURITY

IFIP - The International Federation for Information Processing

IFIP was founded in 1960 under the auspices of UNESCO, following the First World Computer Congress held in Paris the previous year. An umbrella organization for societies working in information processing, IFIP's aim is two-fold: to support information processing within its member countries and to encourage technology transfer to developing nations. As its mission statement clearly states,

IFIP's mission is to be the leading, truly international, apolitical organization which encourages and assists in the development, exploitation and application of information technology for the benefit of all people.

IFIP is a non-profitmaking organization, run almost solely by 2500 volunteers. It operates through a number of technical committees, which organize events and publications. IFIP's events range from an international congress to local seminars, but the most important are:

- The IFIP World Computer Congress, held every second year;
- open conferences;
- working conferences.

The flagship event is the IFIP World Computer Congress, at which both invited and contributed papers are presented. Contributed papers are rigorously refereed and the rejection rate is high.

As with the Congress, participation in the open conferences is open to all and papers may be invited or submitted. Again, submitted papers are stringently refereed.

The working conferences are structured differently. They are usually run by a working group and attendance is small and by invitation only. Their purpose is to create an atmosphere conducive to innovation and development. Refereeing is less rigorous and papers are subjected to extensive group discussion.

Publications arising from IFIP events vary. The papers presented at the IFIP World Computer Congress and at open conferences are published as conference proceedings, while the results of the working conferences are often published as collections of selected and edited papers.

Any national society whose primary activity is in information may apply to become a full member of IFIP, although full membership is restricted to one society per country. Full members are entitled to vote at the annual General Assembly, National societies preferring a less committed involvement may apply for associate or corresponding membership. Associate members enjoy the same benefits as full members, but without voting rights. Corresponding members are not represented in IFIP bodies. Affiliated membership is open to non-national societies, and individual and honorary membership schemes are also offered.

ADVANCED COMMUNICATIONS AND MULTIMEDIA SECURITY

IFIP TC6 / TC11 Sixth Joint Working Conference on Communications and Multimedia Security
September 26–27, 2002, Portorož, Slovenia

Edited by

Borka Jerman-Blažič
Tomaž Klobučar
Institut "Jožef Stefan"
Slovenia

KLUWER ACADEMIC PUBLISHERS
BOSTON / DORDRECHT / LONDON

Distributors for North, Central and South America:
Kluwer Academic Publishers
101 Philip Drive
Assinippi Park
Norwell, Massachusetts 02061 USA
Telephone (781) 871-6600
Fax (781) 681-9045
E-Mail <kluwer@wkap.com>

Distributors for all other countries:
Kluwer Academic Publishers Group
Post Office Box 322
3300 AH Dordrecht, THE NETHERLANDS
Telephone 31 786 576 000
Fax 31 786 576 474
E-Mail <services@wkap.nl>

 Electronic Services <http://www.wkap.nl>

Library of Congress Cataloging-in-Publication Data

A C.I.P. Catalogue record for this book is available from the Library of Congress.

Advanced Communications and Multimedia Security
Edited by Borka Jerman-Blažič and Tomaž Klobučar
1-4020-7206-6

Printed on acid-free paper.
Printed in the United States of America.

Contents

vi

Preface

Security, trust and confidence can certainly be considered as the most important parts of the Information society. Being protected when working, learning, shopping or doing any kind of e-commerce is of great value to citizens, students, business people, employees and employers. Commercial companies and their clients want to do business over Internet in a secure way, business managers when having meetings by videoconferencing tools require the exchanged information to be protected, publishing industry is concerned with the protection of copyright, hospital patients have a right to privacy etc. There is no area in the Information society that can proliferate without extensive use of services that provide satisfactory protection and privacy of data or personality.

In order to gather and present the latest development in the area of communications and multimedia security, and identify future security related research challenges, a Communications and Multimedia Security Conference (CMS 2002) was organised in Portorož, Slovenia, on 26th and 27th of September, 2002. CMS 2002 is the sixth IFIP working conference on communications and multimedia security since 1995. State-of-the-art issues as well as practical experiences and new trends in the areas were the topics of interest again, as proven by preceding conferences.

The book "Advanced Communications and Multimedia Security" contains 22 articles that were selected by the conference programme committee for presentation at CMS 2002. The articles address advanced concepts of communications and multimedia security, such as cryptography, applied cryptography, biometry, communication systems security, multimedia security, digital watermarking, distributed systems security, applications security, and digital signatures. We would like to express our deep appreciation to all authors for their high-quality contributions. Special thanks also go to members of the programme committee:

- Augusto Casaca, INESC, chairman IFIP TC6, Portugal
- David Chadwick, University of Salford, UK
- Bart de Decker, Katholieke Universiteit Leuven, Belgium
- Yves Deswarte, LAAS CNRS, France
- Dieter Gollmann, Microsoft Research, UK
- Ruediger Grimm, TU Ilmenau, Germany
- Patrick Horster, Universitaet Klagenfurt, Austria
- Steve Kent, BBN, USA
- Klaus Keus, BSI, Germany
- Herbert Leitold, IAIK, Austria
- Peter Lipp, IAIK, Austria
- Antonio Lioy, Politecnico di Torino, Italy
- Guenther Pernul, University of Essen, Germany
- Bart Preneel, Katholieke Universiteit Leuven, Belgium
- Fabien A. P. Petitcolas, Microsoft Research, UK
- Wolfgang Schneider, SIT Fraunhofer Gesellschaft, Germany
- Leon Strous, De Nederlandsche Bank, chairman IFIP TC11, Netherlands

Borka Jerman-Blažič and Tomaž Klobučar

ON THE SECURITY OF A STRUCTURAL PROVEN SIGNER ORDERING MULTISIGNATURE SCHEME*

Chris J. Mitchell and Namhyun Hur
Mobile VCE Research Group, Information Security Group
Royal Holloway, University of London
Egham, Surrey TW20 0EX, UK
C.Mitchell@rhul.ac.uk, Namhyun.Hur@rhul.ac.uk

Abstract   Certain undesirable features are identified in the 'Structural proven signer ordering' multisignature scheme of Kotzanikolaou, Burmester and Chrissikopoulos. This scheme is a modification of a previous multisignature scheme due to Mitomi and Miyaji.

Keywords: mobile agent, digital signature, multisignature, cryptanalysis, security

1.   INTRODUCTION

The notion of a multisignature scheme was introduced nearly 20 years ago [Itakura and Nakamura, 1983], and a number of schemes have been proposed since that time. The fundamental idea of a multisignature scheme is that it enables a number of users to collectively create a digital signature on a document (using their own private keys). Typically, all users will sign the same document, and either the order in which they sign will be fixed or, if it is not fixed, then the verifier will not be able to determine in which order the various users signed the document.

For further details on such multisignature techniques, and also on the ElGamal signature scheme on which the cryptosystems described in this paper are based, see, for example, [Menezes et al., 1997].

---

*The work reported in this paper has formed part of the Software Based Systems work area of the Core 2 Research Programme of the Virtual Centre of Excellence in Mobile & Personal Communications, Mobile VCE, www.mobilevce.com, whose funding support, including that of EPSRC, is gratefully acknowledged. More detailed technical reports on this research are available to Industrial Members of Mobile VCE.

1.1   Mitomi-Miyaji multisignatures

Recent papers [Mitomi and Miyaji, 2000, Mitomi and Miyaji, 2001] extend the notion of a multisignature. They provide a model for a multisignature scheme that allows three key properties:

- *message flexibility*, i.e., each party can sign a different document,

- *order flexibility*, i.e., the order in which the various parties create their contribution to the multisignature is not fixed, and

- *order verifiability*, i.e., the order in which the various parties created their contribution to the multisignature can be verified by the verifier of the multisignature.

Mitomi and Miyaji also propose two different multisignature schemes fitting this model, one discrete logarithm based and the other RSA based.

1.2   Multisignatures for mobile agents

In [Kotzanikolaou et al., 2001], the application of Mitomi-Miyaji multisignatures to a mobile agent environment is considered. Specifically, mobile agents (essentially autonomous pieces of code) may visit a number of host platforms, and may wish to collectively sign a message, e.g. to commit to a transaction on behalf of the original sponsor of the agents. Each agent will be equipped with its own (multi)signature private key.

The reason to employ such a model is that single agents may not be trusted to complete a transaction on behalf of a remote sponsor, since their operation may be interfered with by the platform on which they run. In general, there are a number of ways in which the threat posed by a small number of malicious platforms can be reduced. One such approach is to send multiple copies of a transaction agent to a number of platforms, and require that a certain number of copies of the agent (running on different platforms) all consent before the transaction is completed. Each copy of the agent is equipped with a distinct signature key pair (thus preventing an agent on one platform masquerading as an agent executing on a different platform). Of course such an approach requires some co-ordination amongst the various platforms involved, but this is not an issue we consider further here.

A variant of the above approach motivates the particular application of multisignatures we consider here. The model discussed in [Kotzanikolaou et al., 2001] involves a series of agents: $U_1, U_2, \ldots, U_n$ each contributing to a multisignature in turn. Each agent U_i adds its own message string m_i to the evolving multisignature, and thus user U_i actually contributes to a multisignature on a sequence of messages $m_1, m_2, \ldots, m_i$.

We suppose that the recipient of the multisignature will only accept it if a minimum number of distinct agents have contributed to the signature, and that all the agent messages m_i are 'consistent' in some application-specific way.

In this context, [Kotzanikolaou et al., 2001] identify a potential problem with use of Mitomi-Miyaji multisignatures. Specifically, a malicious user can delete one or more of the most recent agent contributions from a multisignature (Kotzanikolaou et al. call this an *exclude* attack). Kotzanikolaou et al. propose two different ways of addressing this problem.

- The first approach, described in Section 3.4 of [Kotzanikolaou et al., 2001], is called a 'simple solution'. It requires signing agent U_j to include in message m_j the identity of U_{j+1}, the agent which U_j selects to be the next entity to contribute to the multisignature. This clearly prevents a malicious party from 'winding back' a multisignature. No changes to the Mitomi-Miyaji schemes are required.

- The second method, described in Section 3.5 of [Kotzanikolaou et al., 2001], is called 'structural proven signer ordering'. This solution actually involves a minor modification to the discrete logarithm based Mitomi-Miyaji scheme. The multisignature computation performed by U_j is modified to include the value of the public key of the next party to the multisignature, namely U_{j+1}. This is designed to achieve the same objective as the simple solution.

  Unfortunately, as we describe below, it is precisely this small modification that enables the manipulation of multisignatures in certain special circumstances. The main conclusion of this paper is therefore that the 'simple solution' is probably preferable.

Specifically, in the remainder of this paper we describe two undesirable features of the structural proven signer modification to the Mitomi-Miyaji discrete logarithm based multisignature scheme.

1.3   Notation and assumptions

We use the notation of [Kotzanikolaou et al., 2001]. Specifically, we suppose that a multisignature is being computed by a series of signers $U_1, U_2, \ldots, U_j$. The part multisignature output by user U_j consists of two sequences of values, namely the messages $m_1, m_2, \ldots, m_j$ (where m_i is chosen by U_i, $1 \leq i \leq j$), and the multisignature components $s_1, s_2, \ldots, s_j$ (where s_i is computed by U_i, $1 \leq i \leq j$), together with the single value r_j.

As in the scheme described in Section 3.5 of [Kotzanikolaou et al., 2001], we suppose that p and g are universally agreed domain parameters, where p is a large prime satisfying $p = 2q + 1$, q is also prime, and g $(1 < g < p)$ has multiplicative order q modulo p.

2. A (PARTIAL) MESSAGE MANIPULATION ATTACK

Suppose a malicious user has succeeded in obtaining iq as its public key, for some integer i. Of course, in general, the malicious user will not know the private key for this public key, i.e. the malicious user will not know a value x for which $g^x \bmod p = iq$. However, this does not prevent at least a partial attack, as we now describe.

2.1   The partial attack

Suppose that a multisignature is being constructed (using the method in Section 3.5 of [Kotzanikolaou et al., 2001]) by a series of signers $U_1, U_2, \ldots, U_j$, and that the next signer (U_{j+1}) is the malicious user; hence U_{j+1} has $y_{j+1} = iq$ as its public key. For convenience we also suppose that $j > 1$, although the attack will work in almost exactly the same way if $j = 1$.

Using the notation of [Kotzanikolaou et al., 2001], U_j will compute

$$
\begin{aligned}
R_j &= g^{k_j} \bmod p, \\
r_j &= (h(m_j\|\mathrm{ID}_j) \cdot r_{j-1})^{-1} \cdot R_j \bmod q, \quad \text{and} \\
s_j &= (x_j r_j + y_{j+1}) \cdot k_j^{-1} \bmod q
\end{aligned}
$$

where x_j is the private key of U_j, h is a hash-function, and y_{j+1} is the public key of user U_{j+1}. Hence, since we know that $y_{j+1} \bmod q = 0$, we have

$$
s_j = x_j r_j k_j^{-1} \bmod q.
$$

User U_j then sends r_j, s_j and m_j to U_{j+1} (together with various other values not of relevance here).

User U_{j+1} can now change the message m_j which user U_j signed. Specifically, suppose user U_{j+1} wishes to make it look as though user U_j signed message $m_j' \neq m_j$. User U_{j+1} first computes $h(m_j'\|\mathrm{ID}_j)$ and then computes

$$
r_j' = r_j \cdot h(m_j\|\mathrm{ID}_j) \cdot (h(m_j'\|\mathrm{ID}_j))^{-1} \bmod q.
$$

This requires no special knowledge. However, the fact that $y_{j+1} \bmod q = 0$ enables U_{j+1} to compute the 'matching' value s_j' using

$$
s_j' = s_j r_j^{-1} r_j' \bmod q = x_j r_j' k_j^{-1} \bmod q = (x_j r_j' + y_{j+1}) \cdot k_j^{-1} \bmod q.
$$

These new values r'_j and s'_j can now be used to replace r_j and s_j in the (partial) multisignature, at the same time that m'_j replaces m_j.

2.2 Completing the attack

Whether or not the process described above is a serious attack depends on whether or not U_{j+1} is in a position to complete the modified multisignature. This depends on whether U_{j+1} possesses the private key x_{j+1} corresponding to the public key $y_{j+1} = iq$. In general this appears to be difficult to arrange.

However, there is one specific case where it is possible for a malicious user to calculate the private key corresponding to a public key congruent to zero modulo q. Suppose, as is often described, the domain parameters p and g are selected as follows.

1 p is chosen so that $q = (p-1)/2$ is prime, and thus precisely $q-1$ of the $p-1$ non-zero elements modulo p, i.e. approximately 50%, will be primitive (see, for example, Section 4.6.1 of [Menezes et al., 1997]).

2 A primitive element modulo p is chosen; call this value e.

3 g is set equal to e^2, guaranteeing that g has order q.

Suppose moreover that $e = 2$. This is not unlikely to be the case; heuristically we expect 2 to be primitive roughly half the time, since roughly half the non-zero elements are primitive, and 2 is typically the first value chosen in a search for a primitive element. In such a case we have $g = 2^2 \bmod p = 4$.

Next observe that $2^q \bmod p = p - 1 = 2q$, and hence $2^{q-1} \bmod p = q$. Thus, $g^{(q-1)/2} \bmod p = 2^{q-1} \bmod p = q$. That is, the private key corresponding to the public key q is simply $(q-1)/2$. Hence, in this special case, if the malicious user chooses his/her public key to be q, then he/she will know his/her own private key, and hence would be able to complete the forged partial multisignature. This represents a serious compromise of the security of the scheme.

Of course, if this particular special case is avoided then the partial signature cannot be completed and the 'partial attack' is simply a (probably unexploitable) questionable property of the scheme.

Finally note that there is one other way in which the above situation can arise. Suppose that, after selecting p (and hence q), g is found by successively examining values 2, 3, 4, and so on, until an element of order q is found. This is a reasonable approach, since small values of g have implementation advantages. Suppose also that 2 and 3 are primitive (and hence are not suitable) — as previously, using heuristic arguments we expect this to be true roughly 25% of the time. Then 4 will have order q and will be selected — exactly the same situation now arises.

3.   A DESTINATION MANIPULATION ATTACK

We show how three different users can conspire to manipulate a contribution to a multisignature made by an honest user.

Suppose that a multisignature is being constructed (using the method in Section 3.5 of [Kotzanikolaou et al., 2001]) by a series of signers $U_1, U_2, \ldots, U_j$, where $j > 2$.

Then, using the notation of [Kotzanikolaou et al., 2001], U_{j-1} will compute

$$\begin{aligned} R_{j-1} &= g^{k_{j-1}} \bmod p, \\ r_{j-1} &= (h(m_{j-1} \| \mathrm{ID}_{j-1}) \cdot r_{j-2})^{-1} \cdot R_{j-1} \bmod q, \quad \text{and} \\ s_{j-1} &= (x_{j-1} r_{j-1} + y_j) \cdot k_{j-1}^{-1} \bmod q \end{aligned}$$

where x_{j-1} is the private key of U_{j-1}, h is a hash-function, and y_j is the public key of user U_j. User U_{j-1} then sends r_{j-1}, s_{j-1} and m_{j-1} to U_j (together with various other values not of relevance here).

Similarly, U_j will compute

$$\begin{aligned} R_j &= g^{k_j} \bmod p, \\ r_j &= (h(m_j \| \mathrm{ID}_j) \cdot r_{j-1})^{-1} \cdot R_j \bmod q, \quad \text{and} \\ s_j &= (x_j r_j + y_{j+1}) \cdot k_j^{-1} \bmod q \end{aligned}$$

where x_j is the private key of U_j and y_{j+1} is the public key of user U_{j+1}. User U_j then sends r_j, s_j and m_j to U_{j+1} (together with various other values not of relevance here).

We now show how a collaboration of three users, namely U_{j-1}, U_{j+1} and a third user which we denote by U'_{j+1}, can modify the multisignature contribution of user U_j to make it look as though the next user specified by U_j was U'_{j+1} and not U_{j+1}. The modifications required are as follows.

First, when computing the original values of R_{j-1}, r_{j-1} and s_{j-1}, user U_{j-1} must choose k_{j-1} equal to x'_{j+1}, where x'_{j+1} is the private key of user U'_{j+1} (we also denote the private key of user U_{j+1} by x_{j+1}). Hence

$$R_{j-1} = g^{k_{j-1}} \bmod p = g^{x'_{j+1}} \bmod p = y'_{j+1}.$$

Second, the values R_{j-1}, r_{j-1} and s_{j-1} are replaced with new values R'_{j-1}, r'_{j-1} and s'_{j-1} computed using a new 'random value' k'_{j-1}, where $k'_{j-1} = x_{j+1}$, the private key of user U_{j+1}.

The replacement values are now computed as follows:

$$R'_{j-1} = g^{k'_{j-1}} \bmod p = g^{x_{j+1}} \bmod p = y_{j+1},$$

$$r'_{j-1} = r_{j-1} \cdot (y'_{j+1} \bmod q)^{-1} \cdot (y_{j+1} \bmod q) \bmod q$$

$$= r_{j-1} \cdot (R_{j-1} \bmod q)^{-1} \cdot (R'_{j-1} \bmod q) \bmod q$$

$$= (h(m_{j-1}\|\mathrm{ID}_{j-1}) \cdot r_{j-2})^{-1} \cdot R'_{j-1} \bmod q,$$

$$s'_{j-1} = (x_{j-1}r'_{j-1} + y_j) \cdot (k'_{j-1})^{-1} \bmod q$$

$$= (x_{j-1}r'_{j-1} + y_j) \cdot (x_{j+1})^{-1} \bmod q.$$

(Note that computing these replacement values is simple since U_{j-1} is a member of the conspiracy).

Replacement values are also computed for R_j, r_j and s_j as follows, this time *without* the co-operation of user U_j:

$$R'_j = R_j, \quad (k_j \text{ is thus as before}),$$

$$r'_j = r_j \cdot (r'_{j-1})^{-1} \cdot r_{j-1} \bmod q$$

$$= (h(m_j\|\mathrm{ID}_j) \cdot r'_{j-1})^{-1} \cdot R'_j \bmod q,$$

$$s'_j = s_j \cdot r'_j \cdot (r_j)^{-1} \bmod q.$$

It remains to show that s'_j has the required properties. Observe that

$$s'_j = s_j \cdot r'_j \cdot (r_j)^{-1} \bmod q,$$

$$= (x_j r_j + y_{j+1}) \cdot k_j^{-1} \cdot r'_j \cdot (r_j)^{-1} \bmod q, \quad \text{(by definition of } s_j\text{)},$$

$$= (x_j r'_j + y_{j+1} \cdot r'_j \cdot (r_j)^{-1}) \cdot k_j^{-1} \bmod q,$$

$$= (x_j r'_j + y_{j+1} \cdot r_{j-1} \cdot (r'_{j-1})^{-1}) \cdot k_j^{-1} \bmod q, \quad \text{(by definition of } r'_j\text{)},$$

$$= (x_j r'_j + y'_{j+1}) \cdot k_j^{-1} \bmod q \quad \text{(by definition of } r'_{j-1}\text{)}.$$

This completes the demonstration, since it is clear that s'_j identifies U'_{j+1} as the next participant in the multisignature instead of U_{j+1}.

4.   ANALYSIS

Observe that, in most circumstances, the (partial) forgery described in Section 2 cannot be completed to a full multisignature. Hence its impact is very limited. Moreover, if users are required to prove possession of their private key before their public key is certified (or otherwise distributed), as is now deemed 'good practice', then in most cases the partial attack is prevented. However, the existence of such a partial attack (which can be extended to a full attack in certain special cases) is nevertheless of concern.

In addition, whilst the forgery described in Section 3 works, and contravenes the required properties of the scheme, it does so in a relatively weak way (given the need for three parties to collaborate to make a small change to the victim's signature). That is, it is hard to see how this attack could be exploited to damage real users of the scheme. However, the existence of such an attack does raise serious questions about the usability of the scheme.

It would therefore appear wise to use the 'simple solution for proven signer ordering' solution, as proposed in Section 3.4 of [Kotzanikolaou et al., 2001], as opposed to the 'structural proven signer ordering' scheme given in Section 3.5 of [Kotzanikolaou et al., 2001]. The use of this former solution has the advantage that it does not change the Mitomi-Miyaji scheme, which has a proof of security, and it is also applicable to any appropriate multisignature scheme.

References

[Itakura and Nakamura, 1983] Itakura, K. and Nakamura, K. (1983). A public-key cryptosystem suitable for digital multisignatures. *NEC J. Res. Dev.*, **71**.

[Kotzanikolaou et al., 2001] Kotzanikolaou, P., Burmester, M., and Chrissikopoulos, V. (2001). Dynamic multi-signatures for secure autonomous agents. In Tjoa, A. and Wagner, R., editors, *DEXA Workshops — Proceedings 12th International Workshop on Database and Expert Systems Applications (DEXA 2001), 3-7 September 2001, Munich, Germany*, pages 587–591. IEEE Computer Society.

[Menezes et al., 1997] Menezes, A., van Oorschot, P., and Vanstone, S. (1997). *Handbook of Applied Cryptography*. CRC Press, Boca Raton.

[Mitomi and Miyaji, 2000] Mitomi, S. and Miyaji, A. (2000). A multisignature scheme with message flexibility, order flexibility and order verifiability. In Dawson, E., Clark, A., and Boyd, C., editors, *Information Security and Privacy: Proceedings of the 5th Australasian Conference — ACISP 2000*, number 1841 in Lecture Notes in Computer Science, pages 298–312. Springer-Verlag, Berlin.

[Mitomi and Miyaji, 2001] Mitomi, S. and Miyaji, A. (2001). A general model of multisignature schemes with message flexibility, order flexibility and order verifiability. *IEICE Trans. Fundamentals*, **E84-A**:2488-2499.

RENEWING CRYPTOGRAPHIC TIMESTAMPS

Sattam S. Al-Riyami and Chris J. Mitchell
Information Security Group, Royal Holloway, University of London
Egham, Surrey TW20 0EX, UK
S.Al-Riyami@rhul.ac.uk, C.Mitchell@rhul.ac.uk

Abstract   This paper shows that the scheme described in Haber and Stornetta [Haber and Stornetta Jr., 1994] for extending the validity of a cryptographic timestamp for a Time Stamping Service contains security shortcomings. A modification is proposed to rectify the identified shortcomings.

Keywords: timestamping, TSA, TSS, digital signature, PKI, security, protocol failure

1.   INTRODUCTION

A time-stamping service (TSS) has been identified by both the IETF and ISO/IEC as a potentially important part of a Public Key Infrastructure (PKI), and draft standards have been produced by both bodies [Adams et al., 2001, ISO/IEC, 2001]. Conventionally, and as originally proposed by Haber and Stornetta in 1991 [Haber and Stornetta, 1991], a TSS will take as input a hash-code of a data string supplied by a client, and will return a digital signature computed on a concatenation of this hash-code and a time-stamp (the *cryptographic timestamp*). This cryptographic timestamp can then be used as evidence that the original data string existed at the time indicated, without revealing the data string to the TSS. The hash-code should be computed using a collision-resistant one-way hash-function (see for example [Menezes et al., 1997]).

One particularly important application of a TSS is to prolong the lifetime of a digital signature. Without use of a TSS, when a public key certificate expires or is revoked, all signatures computed using the corresponding private key potentially lose their validity. This is because, if the private key becomes known, it is possible to forge signed documents that are indistinguishable from documents produced prior to the point at which the private key was compromised. On the other hand, if a TSS adds a signed timestamp to a signed document, then this proves that the signature on the document was created prior to the time at which the timestamp was created. Even if the signing key is subsequently revoked, the fact that the original signature can be shown to have been created prior to the time of revocation means that the signature remains valid. Of course, in some environments, and depending on the policy in force, signatures may not lose their validity if the public key revocation occurs for reasons other than key compromise. However, even in such cases, problems may eventually arise because key lengths deemed secure at the time the signature was created may no longer be deemed secure at some later date.

This is particularly relevant for circumstances where signatures are needed to have long-term validity. One obvious example where this will be the case is for the signing of high-value financial transactions and/or contracts, where, as with handwritten signatures, digital signatures will be expected to last indefinitely. It is therefore very likely that timestamping services will be of particular importance for PKIs used to support security for financial applications. This issue is also discussed in RFC 3161, [Adams et al., 2001, Appendix B].

If the signature key of the TSS itself is about to expire, then it is typically necessary to re-timestamp the original cryptographic timestamp with a new TSS signature key, if the original timestamp is to remain valid. This issue has been widely discussed in the literature (see for example [Bayer et al., 1993]). However, independently of the security of the digital signature created by the TSS, the strength of a cryptographic timestamp also relies on the security of the hash function used to compute the hash-code submitted to the TSS [Preneel et al., 1998]. This paper focuses on ways of dealing with the situation where this hash-function is broken, since such an event has the potential to invalidate all cryptographic timestamps computed with this hash-function.

Other authors discuss witnessing and linking techniques in order to improve the security and accountability of the TSS scheme [Buldas et al., 2000]. However, if a pre-image of the hash-code is discovered at any stage after the cryptographic timestamp is produced, the additional techniques will not provide proof of prior existence for the document. Increases in computing power and new algorithmic techniques mean that current "trusted" hash-functions are likely to eventually need replacement. Hence, as soon as any doubts about future use of a particular hash-function arise, and before it is known to be broken, the cryptographic timestamp should be renewed.

Note that loss of confidence in the hash-function is not the only reason for renewal of a cryptographic timestamp. Indeed, timestamp renewal is more commonly discussed in the context of update of the TSS key pair. However, for the purposes of this paper, renewal of cryptographic timestamps refers only to updating the hash-function.

2. THE HABER-STORNETTA RENEWAL PROTOCOL

Haber and Stornetta considered the hash-function renewal problem very briefly in their original 1991 paper [Haber and Stornetta, 1991], and Bayer et al. proposed some modifications in [Bayer et al., 1993]. A much more concrete proposal appeared in the Haber and Stronetta patent [Haber and Stornetta Jr., 1994], and it is this latter scheme we consider here. We first describe their basic time-stamping protocol, and then go on to outline their proposed solution to dealing with hash-functions that require renewal. Note that the form of the timestamping protocol included in the latest draft standards [Adams et al., 2001, ISO/IEC, 2001] is very similar to the Haber and Stornetta proposal.

The basic timestamping protocol, as described in [Haber and Stornetta Jr., 1994], has two steps. First the client requests a timestamp from the TSS, as follows:

A→TSS: $h(M\|X) = h(R)$,
> where A denotes the client of the TSS, M is the 'message' to be timestamped, $\|$ denotes concatenation of data items, X is other data of unspecified form as chosen by the client, and h is a hash-function. The concatenation of M and X is also written as R (for *receipt*). Note that the Haber and Stornetta patent [Haber and Stornetta Jr., 1994] is not completely clear as to the purpose of the data string X — they simply suggest that it is used to identify M by, for example, including the 'author data'.

Second, the TSS responds with the cryptographic timestamp C:

TSS→A: $C = S_{TSS}(h(R)\|T)$,
> where S_{TSS} denotes the digital signature function of the TSS, and T is the time/date stamp.

On receipt of C, the client A stores it as evidence that M existed at time T.

Note that in their patent specification, [Haber and Stornetta Jr., 1994] Haber and Stornetta actually propose the use of a cryptographic timestamp function F, which is used to compute C, i.e. $C = F(R)$. We have chosen the simplest interpretation of F, namely that F involves concatenation with a timestamp and applying the TSS's digital signature function. In fact, F could also involve the concatenation of data in addition to the timestamp, to support more complex variants of the scheme. However, this is outside the scope of this paper.

The protocol for renewing the cryptographic timestamp to extend its lifetime is as follows:

A→TSS: $h'(C\|M)$,
> where h' denotes the replacement hash-function.

The TSS responds by sending the extended cryptographic timestamp back to the client;

TSS→A: $C' = S'_{TSS}(h'(C\|M)\|T')$,
> where C' is the extended cryptographic timestamp, T' is the time of the renewal request and S'_{TSS} is the new signature function of the TSS.

It is also stated in [Haber and Stornetta Jr., 1994] that an alternative more secure way of obtaining a cryptographic timestamp involves use of a compound cryptographic timestamp. Compound cryptographic timestamps are essentially the same as the previous cryptographic timestamps but use two different trusted hash functions, which are applied in parallel to the same receipt. To initiate the basic protocol the client sends $h(R)\|h'(R)$ and consequently two signatures are produced by the TSS. This potentially increases the lifetime of the cryptographic timestamp before a renewal is required.

3.   ATTACK AND OBSERVATION

We now describe a possible attack against the Haber and Stornetta hash-function renewal protocol. In the attack we suppose that the hash-function used in the original timestamping protocol has been compromised at some point after the renewal process, i.e. the very situation which renewal is supposed to deal with.

The attack is initiated after the first basic cryptographic timestamp is issued and is completed at a later time. In this attack there is no need to break h' and we only assume that the initial hash h is compromised. The attack is performed as follows:

1 Get C: Using the basic protocol, the attacker obtains the cryptographic timestamp $C = S_{TSS}(h(R)\|T)$. Note that we assume $R = (M\|X)$ for some X.

2 Obtain C': The attacker at time T' wants to backdate a forged message M' to T. He does this by requesting a renewal for the original cryptographic timestamp C by sending $h'(C\|M')$ to the TSS, thus obtaining $C' = S'_{TSS}(h'(C\|M')\|T')$. Now at time T', both hash-functions h and h' remain secure, and therefore at this stage C' cannot be used as the basis of an attack on the scheme.

3 Break h: At a later time T'', suppose h has been broken, so that it is no longer a one-way function. That is, we assume that pre-images can be found for h. Suppose that, as a result, the attacker is able to find a pre-image for $h(R)$ under h, i.e. the attacker can find a string $R^\star$ such that $h(R^\star) = h(R)$. Suppose, moreover, that the attacker can choose the first part of the pre-image $R^\star$ — in particular we suppose that the attacker chooses $R^\star$ so that $R^\star = M'\|X^\star$ for some string $X^\star$. We are thus supposing that h has been subject to a particularly severe form of failure, i.e. so that we can find pre-images for which the first part can be freely chosen. Note however that, given that h involves the iterative use of a round-function (as is the case for all commonly used hash-functions), such a catastrophic failure of the one-way property is not unlikely if the round-function is found to be weak.

The attacker can now claim that the message M' was signed at time T, with C as proof and renewed cryptographic timestamp C' as supporting evidence. Before proceeding, observe that, during the attack described above, in order to find the string $R^\star$ such that $h(R^\star) = h(R)$, where $R^\star = M'\|X^\star$ for some string $X^\star$, some part of $X^\star$ will need to be effectively 'random'. It might consequently be argued that this will reveal the attack, since the third party adjudicating in a dispute will observe that $X^\star$ is a meaningless sequence.

However, whilst this may be true in some circumstances, there is a danger that the situation will not always be so clear cut. First, it might be possible to choose the first part of $X^\star$ to conform to what is expected of such strings, and then to arrange for the random part to be disguised (e.g. as random padding of some kind). Second, the patent specification merely states that the string X can be chosen by the client, and hence it is not reasonable to expect a third party adjudicator to decide what was in the client's mind when he created the string X. Third, it is not good practice to design protocols which rely on third parties making judgements about whether a string is meaningful or not. The protocol should be designed to avoid such issues, and we show below how this can be achieved in a very simple way.

It should be clear that the main reason that this attack is possible is that the original timestamp involves signing $h(R)$, where $R = M\|X$, and the timestamp renewal involves signing $h'(C\|M)$. That is the two timestamps actually involve different data strings, namely $M\|X$ in the first case, and M in the second case. It is this difference that allows the attack to take place. This observation motivates the proposed protocol modifications discussed in the next section.

4.   REVISED VERSIONS OF THE PROTOCOL

Based on the above observation on the cause of the attack, we propose that the protocol should be modified in the following minimal way.

To obtain a basic cryptographic timestamp, the client initiates the request and the TSS responds, as follows:

A→TSS: $h(Y)$,
> where Y denotes either M or $R = M\|X$. Whichever version of Y is adopted, it must be preserved exactly in the renewal protocol.

TSS→A: $C = S_{TSS}(h(Y)\|T)$.

The protocol for renewing the cryptographic timestamp in order to extend its lifetime is as follows:

A→TSS: $h'(C\|Y)$,

TSS→A: $C' = S'_{TSS}(h'(C\|Y)\|T')$.

The only change is to fix the fundamental problem which led to the previous attack, namely that the two timestamps involve different bit-string inputs into the hash-function. Note that, in the case $Y = M$, this modification has previously been described in various places, including in [Haber and Stornetta, 1997]. However, it is important to note that the reason to adopt this variant (or the variant with $Y = M\|X$) in favour of the protocol described in [Haber and Stornetta Jr., 1991], i.e. the existence of the attack described above, has never previously been discussed.

We now consider certain other possible modifications to both the basic and the renewal protocols. First note that we are essentially using $h(Y)$ in the basic protocol and $h'(h(Y)\|Y)$ for the renewal request. An equally secure alternative would be to use $h(Y)$ and $h(Y)\|h'(Y)$ in the respective steps. However, there is no added value in this alternative. In fact the scheme will require the TSS to differentiate between new timestamps and renewal ones. This will unnecessarily complicate the practical implementation of the TSS, in addition to using more bandwidth.

A second modification is to include a hash-function identifier with the hash-code [ISO/IEC, 2001]. This is of fundamental importance in ensuring that the hash-code can not be deliberately mis-represented as having been generated using a different (weaker) hash-function.

A third modification is the inclusion of the message length with the hash-code [PKITS, 1998]. This limits the freedom of an attacker attempting to find a second pre-image for the hash-code.

With these latter two modifications, the basic timestamping protocol is now as follows:

A→TSS: $h(Y)\|N\|hID,$
    where hID is a one byte hash-identifier and N is the length of Y.

TSS→A: $C = S_{TSS}(h(Y)\|N\|hID\|T),$

The corresponding modified protocol for renewing a cryptographic timestamp to extend its lifetime is as follows:

A→TSS: $h'(C\|Y)\|N\|hID',$
    where hID' denotes the hash identifier of h'.

TSS→A: $C' = S'_{TSS}(h'(C\|Y)\|N\|hID'\|T').$

5. CONCLUSION

It has been shown that, in the case where the original hash-function admits pre-image attacks, the Haber and Stornetta cryptographic timestamp renewal scheme [Haber and Stornetta Jr., 1994] does not prevent retrospective forgeries. However, this renewal scheme was designed to prevent forgeries in precisely these circumstances. A modification to the protocol that prevents these attacks has been proposed. The modified protocol has a computational and communications overhead that is very similar to the original scheme.

References

[Adams et al., 2001] Adams, C., Cain, P., Pinkas, D., and Zuccherato, R. (2001). *Internet X.509 Public Key Infrastructure, Time Stamp Protocol (TSP)*. Internet Engineering Task Force (IETF). Request For Comments: 3161.

[Bayer et al., 1993] Bayer, D., Haber, S., and Stornetta, W. S. (1993). Improving the efficiency and reliability of digital time stamping. In Capocelli, R., De Santis, A., and Vaccaro, U., editors, *Sequences II: Methods in Communication, Security and Computer Science*, pages 329–334. Springer-Verlag.

[Buldas et al., 2000] Buldas, A., Lipmaa, H., and Schoenmakers, B. (2000). Optimally efficient accountable time-stamping. In Zheng, Y. and Imai, H., editors, *Public Key Cryptography*, volume 1751 of *Lecture Notes in Computer Science*, pages 293–305, Melbourne, Australia. Springer Verlag.

[Haber and Stornetta, 1991] Haber, S. and Stornetta, W. S. (1991). How to time-stamp a digital document. *Journal of Cryptology*, 3:99–111.

[Haber and Stornetta, 1997] Haber, S. and Stornetta, W. S. (1997). Secure names for bit strings. In *Proceedings of the 4th ACM Conference on Computer and Communication Security*, pages 28–35. ACM.

[Haber and Stornetta Jr., 1994] Haber, S. A. and Stornetta Jr., W. S. (1994). *Method of extending the validity of a cryptographic certificate*. Bell Communications Research Inc., Livingston, N.J. United States Patent number 5,373,561.

[ISO/IEC, 2001] ISO/IEC (2001). *ISO/IEC FCD 18014-1, Information technology — Security techniques — Time-stamping services — Part 1: Framework*. International Organization for Standardization, Geneva, Switzerland.

[Menezes et al., 1997] Menezes, A., van Oorshot, P., and Vanstone, S. (1997). *Handbook of Applied Cryptography*. CRC Press, Boca Raton.

[PKITS, 1998] PKITS (1998). *PKITS — Public Key Infrastructure with Time-Stamping Authority*. Fábrica Nacional de Moneda y Timbre (FNMT), Madrid, Spain. ETS Project: 23.192, Architecture of Time-Stamping Service and Scenarios of Use: Services and Features.

[Preneel et al., 1998] Preneel, B., Rompay, B. V., Quisquater, J.-J., Massias, H., and Avila, J. S. (1998). Design of a timestamping system. Technical report, TIMESEC, Katholieke Universiteit Leuven and Université Catholique de Louvain.

IMPLEMENTING ELLIPTIC CURVE CRYPTOGRAPHY

Design of a Standard Compliant Java Library

Wolfgang Bauer
Institute for Applied Information Processing and Communications
Graz University of Technology
wolfgang.bauer@iaik.at

Abstract   Properties like short keys and efficient algorithms make elliptic curve cryptography (ECC) more and more interesting for future oriented applications. In this paper we give a short overview of the basics of ECC. Thereby we show where programmers can gain possible speed-ups and what parts are crucial. Since there are many different implementation options and some of the algorithms are patented, we believe that there is no optimal solution. Therefore we introduce a software framework, which allows a transparent replacement of data and algorithms. Furthermore, we discuss aspects of the standardized encoding, and point out where interoperability problems could occur.

Keywords: ECC, Java

INTRODUCTION

The security of public key cryptography is based on hard mathematical problems. Today, the popular algorithms are either based on the integer factorization problem (IFS), like RSA, the discrete logarithm problem (DLP), like DSA, or the elliptic-curve discrete-logarithm problem (ECDLP). The only problem where no sub-exponential time algorithm is known so far is the ECDLP. That is why the key lengths of elliptic-curve based algorithms are much shorter and increase much slower over time. The Austrian signature ordinance, for instance, stipulates RSA keys of at least 1023 bits, whereas ECDSA keys may be 160 bits only (by the end of the year 2005).

Therefore ECDLP based algorithms getting more and more important especially for smart cards, which are very limited in memory. For this reason the Austrian Social Insurance Smart Card, which every Austrian citizen should have at the beginning of the year 2003, uses the Elliptic Curve DSA (ECDSA) for secure electronic signatures.

In this paper we introduce design aspects and problems of a standard compliant ECC Java library. This library is designed to work not only for signature creation, but also for verification. This means, that it must be able to deal with all requirements defined in the standards. There are numerous of papers which discuss different aspects and algorithms of ECC, whereas this paper focuses on implementing all of the standard requirements. We point out software design decisions and where interoperability problems may occur.

The rest of this paper is organized as follows. The next section provides a short overview of elliptic-curve cryptography. Afterwards we discuss components and implementation options of an ECC software system. Section 3 defines the design goals and section 4 introduces the actual design of an ECC Java library. Finally conclusions are given.

1.   BASICS OF ELLIPTIC CURVE CRYPTOGRAPHY

The idea of ECC was first introduced by Neil Koblitz and Victor Miller in the mid eighties. Since then cryptographers started to analyze the ECDLP and so far no vulnerabilities have been discovered. First ECC was an academic concern but in recent years it started to gain commercial interest, which lead to standards like [7] and [8].

To understand ECC you have to know a lot of math, which is usually scaring for programmers. Therefore one common way is to start with the following graphical example. Figure 1 shows an elliptic curve ($y^2 = x^3 + ax + b$) defined over the field of real numbers ($a, b, x, y \in \mathbb{R}$).

We define a basic operation, the point addition. The sum of two points is determined as shown in figure 1. Draw a line through the points P and Q and the intersection of this line with the curve is the Point -(P+Q). Now take the negative y-coordinate (which again is a point on the curve) and you are done. For doubling a point use the tangent and proceed as with the point addition. This rule obviously will not work if the line through the point(s) is parallel to the y-axis. Therefore we define the point at infinity ($\mathcal{O}$) as a result of this operation.

Calculations with infinite fields, as with $\mathbb{R}$ in the example above, are not suitable for cryptographic purposes. That is why finite fields F_q, as described in section 2.1, are used. The set $E(F_q)$ consists of all points $(x, y), x \in F_q, y \in F_q$ satisfying the elliptic curve equation and the point $\mathcal{O}$.

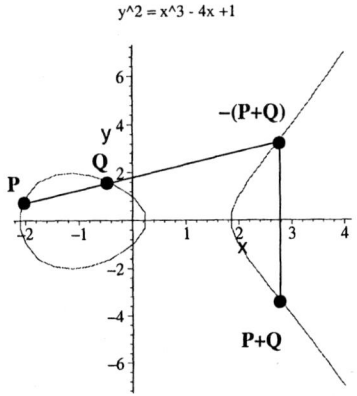

Figure 1. Elliptic Curve Point Addition

Table 1. Point Addition and Doubling Formulas
$P = (x_1, y_1)$ *and* $Q = (x_2, y_2) \in E(F_q)$; $P \neq Q$

$P + Q = (x_3, y_3)$	$x_3 = \left(\frac{y_2 - y_1}{x_2 - x_1} \right)^2 - x_1 - x_2$
	$y_3 = \left(\frac{y_2 - y_1}{x_2 - x_1} \right)(x_1 - x_3) - y_1$
$2P = (x_3, y_3)$	$x_3 = \left(\frac{3x_1^2 + a}{2y_1} \right)^2 - 2x_1$
	$y_3 = \left(\frac{3x_1^2 + a}{2y_1} \right)(x_1 - x_3) - y_1$

Table 1 shows the formulas of the previously described point operations. It can be shown that $E(F_q)$ together with the point addition forms a group. We write the "+" for the group operation (point addition) and the neutral element is $\mathcal{O}$. Thus $\mathcal{O} + P = P + \mathcal{O} = P \; \forall \; P$ of $E(F_q)$.

With this additive group [9] defines the ECDLP problem as follows. Given an elliptic curve E defined over a finite field F_q, a Point $P \in E(F_q)$ of order n, and a point $Q = lP$ where $0 \leq l \leq n - 1$, determine l.

Based on this hard mathematical problem one can adapt algorithms based on the DLP to ECDLP by replacing $\mathbb{Z}_p^*$ with the elliptic curve group. After this short introduction to the EC arithmetic, the next section shows the basic components of a software implementation.

2.   COMPONENTS AND IMPLEMENTATION OPTIONS

There are many other papers, like [5] [6] [1], comparing and explaining aspects of elliptic curve crypto systems. Therefore this section only points out where software developers have various implementation options and what has to be heeded. According to our software design, introduced in section 4, this section is subdivided into finite field arithmetic, EC arithmetic, and data formatting options.

2.1   Finite Field Arithmetic

Finite fields form the basis of EC arithmetic. As the formulas of table 1 show, one simple point addition requires many operations in the underlying finite field. Therefore special attention and efficient algorithms are required.

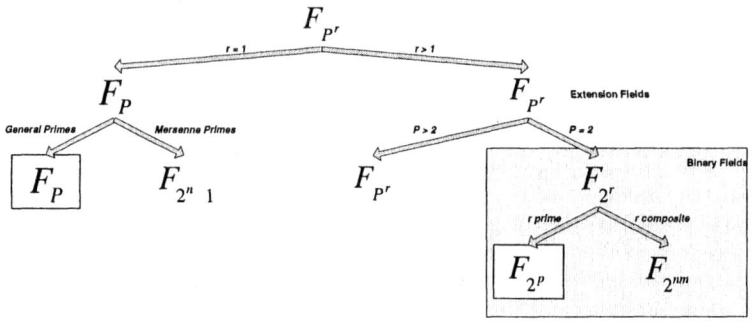

Figure 2.   Classification of Finite Fields

Figure 2 shows a classification of finite fields. The two boxed leafs of this tree (F_P and F_{2^p}) are used for ECC and are specified in the standards.

2.1.1 Prime Fields.

Prime fields are the most efficient choice for software implementations. Many papers like[6] describe algorithms for long integer arithmetic. The basic addition and multiplication operations use conventional integer arithmetic based on the processor's word size. Therefore they are very fast.

The crucial operations are the reduction modulo a prime and the calculation of the multiplicative inverse element. By the use of projective coordinates the inversion in the formulas of table 1 can be avoided. Only one inversion at the end of the ECC calculation is needed to regain the affine coordinates. Therefore the costs for this operation can nearly be neglected. The major design decision in prime field arithmetic is how to integrate the modular reduction and how to represent field elements (e.g. use a Montgomery or Barret reduction).

As a last point we should mention, that in general the square operation can be done more efficient than a multiplication. This is the reason why we decided to put this operation into the list of basic field operations, as show in section 4.

2.1.2 Binary Fields.

There are several ways of representing field elements in F_{2^p}, whereby polynomial basis representation seems to be the most efficient [6]. Unfortunately, to be standard compliant, one must be able to deal with Gaussian Normal Basis (GNB) representations as well. Since almost every algorithm in GNB seems to be patented and this representation does not provide any benefits, we decided not to support it at all. Nevertheless, we still have the option to implement a base transformation and therewith gain standard compliance subsequently.

In polynomial base representation the field elements are considered as polynomials with coefficients $\in F_2$. Addition, which is the same as subtraction, is a simple XOR operation. Multiplication is a conventional polynomial multiplication modulo an irreducible polynomial.

Microprocessors do not support polynomial arithmetic and therefore software implementations of F_{2^p} multiplication are usually slow. However, modulo reduction can be performed rather efficient. That is especially true for trinomials and pentamonials, which are the standardized reduction polynomials. In binary fields the programmer has to consider what algorithms to use and data formats that allow efficient polynomial arithmetic. As in prime fields squaring can be done more efficient.

2.2 EC Arithmetic

Efficient finite field arithmetic is the key for a high performance ECC implementation. Even so the EC group arithmetic offers a lot of tuning possibilities. The basic operation for ECC is the computation of the scalar product $k * P$ $0 \leq k \leq n - 1$, which is a repeatedly application of the basic group operation. Depending on the field type (prime or binary) the formulas are a little bit different. The actual group for the calculation is determined by a set of so called domain parameters. It consists of the following items.

- The field size, which defines the underlying field F_q. In case of binary fields the used basis type and in case of a polynomial base representation the irreducible polynomial, is required.

- The Curve Parameters a and b $\in F_q$

- The Base Point G $= (x_G, y_G)$, $x_G, y_G \in F_q$ on E of prime order.

- The Order n (of the base point G).

- The cofactor h $= \#E(F_q)/n$

- Optional parameters.

Depending on the actual domain parameters EC arithmetic offers some implementation variants. For instance there exist some elliptic curves, where more efficient algorithms are known (e.g. Koblitz curves). Further speed-ups can be achieved by the fact that in elliptic curve groups one gets the negative element nearly for free. This allows the construction of efficient addition-subtraction chains for the scalar multiplication [4]. The point representation influences the overall performance as well, and the type of coordinates (affine, projective or mixed) to be used, has to be considered.

2.3 Algorithms

All the previously mentioned parts form the basis for an elliptic curve crypto system. Algorithms originally based on the discrete logarithm problem can be adapted to work on elliptic curves. Though their exist other ECC algorithms like an EC version of the Diffie-Hellman key-agreement scheme or encryption schemes, the most famous EC algorithm is the ECDSA. It can be subdivided in 3 steps.

1 Message Digesting: using SHA-1

2 EC Computation: calculation of k*G (k randomly chosen)

3 Modular Computations: modulo n

The third point always requires some arithmetic modulo a prime. Therefore it is not possible to calculate an ECDSA signature with binary field arithmetic only. But this fact is mainly important for signature creation devices.

2.4   Data Formatting and Standards

The sections above have shown how elliptic curve crypto systems operate. To use them in praxis and to ensure interoperability one has to present the data in a system independent and standartized way. Several ECC standards exist and figure 3 shows their compatibility.

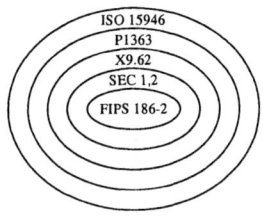

Figure 3.   Compatibility of ECC Standards

To make sure an ECDSA signature can be verified, the other party must know the public key and the related domain parameters. Therefore the standards define structures for the encoding of all the data. There are two drawbacks of this approach.

- Memory requirements: the size of the encoded domain parameters is much larger than the signature and the keys.

- The verifying party must either rely on the validity of the domain parameters or perform some checks.

To circumvent this unsatisfying situation the possibility exists to use a unique object identifier. This approach will only work if the domain parameters are standardized. [7] [8] and [3] altogether define OIDs for 20 curves over prime fields (112 - 521 bits) and for 38 curves over binary fields (113 - 571 bits). We believe that this variant is the best, since there is no argument to use other curves than these standardized. Application do not have to do a parameter validation and the encoding size is negligible. One disadvantage is, that every party must have a table, assigning OIDs to domain parameters.

There is a third variant, where domain parameters are omitted at all. The application either implicitly knows them, or within a public key infrastructure, these domain parameters may be inherited from a Certification Authority. This approach might be useful in some special cases, but in general it could cause interoperability troubles.

Patented algorithms might cause further interoperability problems. Unfortunately some of them are contained in the standards. For instance the following three ways of encoding elliptic curve points are standardized.

- The uncompressed form, which consists of the x and y coordinate of this point.

- The compressed representation only contains the x coordinate and one single bit indicating which of the two possible points to take. To decompress a point, one has to solve the elliptic curve equation with a given x value, which is a quadratic equation.

- A hybrid form, which is a mixture of the points above.

Certicom holds a patent on the point compression and thus not every software might support this feature. Nevertheless a good software design allows all of the encoding possibilities above. To take patent issues into consideration the design should support plug-able modules and easy configuration.

A further point that should be considered is the encoding of the data structures, like domain parameters and public keys. Until today the most common format is the Abstract Syntax Notation 1 ASN.1. The ECC standards define such ASN.1 structures und thereby allow ECDSA to be used with existing X.509 certificates (private key encoding for use in PKCS#8 is only defined in [2]). As an alternative the Extensible Markup Language XML is getting more and more important nowadays. But the syntax for ECDSA with XML signatures is only a draft yet. Nevertheless software designers should keep in mind that other formatting options may be desired in the future.

3.   DESIGN GOALS

As already stated in the introduction, the software design is very much dependent on the actual deployment. For instance, software for signature creation devices usually has to implement only a small part of the corresponding standard. That means one can optimize code for one set of domain parameters (see section 2) and choose one of the encoding possibilities. Furthermore, the signature creation devices may leave some tasks to the environment. Memory and computational limited devices may just compute the signature value and the formatting and encoding may be left to the application.

In this paper we focus on the more general case, which deals with both, signature creation and verification. The challenge is to write code, which provides for integration of all standardized features. Since there are many implementation options and design tradeoffs there is not one best way. Furthermore, new faster algorithms might pop up and subsequent integration should not be problem. Finally, patent issues have to be considered and thus algorithms and data representations should be pluggable. The main points of our design guidelines can be summarized with the following key words.

- Modularity: Subdivide the problem and make the modules independent.

- Extensibility: Easy integration of new algorithms and data representations.

- Maintainability: The code should be well structured and easy to understand.

- Performance: At first view a large software framework (with abstract classes and interfaces) and high performance seem to be contradictory design goals. However performance is primary influenced by the used algorithms and therefore a flexible and clear design pays off.

- Robustness: Ensure a well defined behavior particularly in case of an error.

After this definition of the design goals, the next section introduces the software framework. We will pick out some parts and explain how some of the points above can be achieved.

4. SOFTWAREDESIGN

The design of the whole library would go beyond the scope of this paper. Therefore this section introduces the overall module concept and explains some features in more detail.

The main parts of the library are shown in Figure 4. According to the design criteria above, each module operates independent. The behavior of the modules is defined by interfaces and the actual implementation is not visible from outside. This design allows a transparent replacement of the algorithms and data representations.

Finite Field Arithmetic. The common interface for all finite fields defines the basic field operations like:

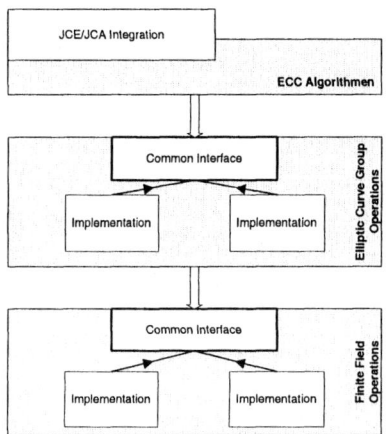

Figure 4.   Software Module Design

```
public interface Field {
  public FieldElement getONEelement();
  public FieldElement getZEROelement();
  public void negate(FieldElement a);
  public void add(FieldElement a, FieldElement b);
  public void invert(FieldElement a);
  public void multiply(FieldElement a, FieldElement b);
  public void square(FieldElement a);
  ...
}
```

As you can see, there are no implementation specific elements within this interface. The square operation, which is not really required, was added for performance reasons. The interface contains some additionally methods to get the field type and some other utility functions that are not listed above. Furthermore, the finite field module contains interfaces for field elements and their values. Thereby implementations can choose any desired data structures and algorithms.

Elliptic Curve Arithmetic.   The structure of this module is very similar to the previously presented. Again the developer has the complete freedom of how to present the data and which algorithms to use. Depending on patent issues it is possible to change algorithms without any modification in the rest of the whole library. Also the coordinate types (affine, projective or mixed) are not hard coded and can be changed transparently. The common interface offers methods for the basic group operation. To speed up the ECDSA signature verification process we decided to put a method for simultaneous multiplications ($P = h_1 G + h_2 W$) into the interface (see [10]).

Elliptic Curve Algorithms. This module contains the ECC algorithms like ECDSA or ECDH. It defines interfaces for private and public keys. Furthermore the administration of domain parameters and encoding specific tasks are placed here. Applications access the library through the Application Programming Interface API. The next session discusses the integration of this API with the Java framework.

Java Integration. Security has been one of the main design goals of the Java programming language. Therefore it offers a large framework for cryptographic algorithms called Java Crypto Extension (JCE)/Java Crypto Architecture (JCA). The so-called provider concept allows third party implementation of various algorithms but until now no framework for ECC exists. One can use the existing signature interface to work with ECDSA but there is no common way to specify the domain parameters and keys. Therefore, always some proprietary code must be included in the application.

A further shortcoming of the Java framework is the design of the BigInteger class. This class is usually used for arbitrary long integer arithmetic. Unfortunately it has two serious disadvantages for the use in ECC.

- Instances of this class (objects) are immutable. This means, for every finite field operation a new object must be created. Since ECC computations require many operations in the underlying finite field, many of the BigInteger objects have to be created, which is not very efficient.

- There is no public method for the often used square operation and therefore the slower multiply method has to be taken.

For the reasons above, it makes sense to implement an own, mutable BigInteger class.

5. CONCLUSIONS

Short keys and efficient algorithms make ECC interesting for future oriented applications. This paper provides an introduction into ECC. We have discussed various parts of a standard compliant software library, like finite field and EC arithmetic. The numerous implementation variants and encoding options, as well as possible patent issues, require a flexible software architecture. We have introduced a design, which enables programmers to transparently add, remove, or replace algorithms and data types. This framework might be considered as a performance penalty. Usually the main speed-ups can be achieved by efficient algorithms and data structures and therefore the advantages of a plug-able design pay-off. Furthermore, a clear and modular structure provides for easy integration of new algorithms guarantees the maintainability.

References

[1] Michael Brown, Darrel Hankerson, Julio Lopez, and Alfred Menezes. Software implementation of the NIST elliptic curves over prime fields. In *CT-RSA*, pages 250–265, 2001.

[2] Certicom. Sec 1: Elliptic curve cryptography, 2000.

[3] Certicom. Sec 2: Recommended elliptic curve domain parameters, 2000.

[4] Daniel M. Gordon. A survey of fast exponentiation methods. *J. Algorithms*, 27(1):129–146, 1998.

[5] Jorge Guajardo and Christof Paar. Efficient algorithms for elliptic curve cryptosystems. In *CRYPTO*, pages 342–356, 1997.

[6] Darrel Hankerson, Julio Lopez Hernandez, and Alfred Menezes. Software implementation of elliptic curve cryptography over binary fields. In *Cryptographic Hardware and Embedded Systems*, pages 1–24, 2000.

[7] IEEE. 1363 standard specification for public key cryptography, 2000.

[8] American National Standards Institute. X9.62-1998, public key cryptography for the financial services industries: The elliptic curve digital signature algorithm (ecdsa), 1998.

[9] D. Johnson and A. Menezes. The elliptic curve digital signature algorithm (ecdsa), 1999.

[10] Bodo Moeller. *Algorithms for Multi-exponentiation*, pages 165–180. Springer-Verlag, 2001.

A NEW ASYMMETRIC FINGERPRINTING FRAMEWORK BASED ON SECRET SHARING

Yan Wang, Shuwang Lü and Zhenhua Liu
State Key Laboratory of Information Security,
Chinese Academy of Science, Beijing, China *
ywang_cas@yahoo.com

Abstract   This paper presents a new type of asymmetric fingerprinting framework based on the idea of secret sharing. With the help of FIC(Fingerprint Issuing Center), we achieve an efficiency improvement in implementing asymmetric fingerprinting protocol. It avoids using secure multi-party computations. Further research directions are also suggested.

Keywords: Copyright protection, Asymmetric fingerprinting, Secret sharing

1.   INTRODUCTION

Recent years have seen a rapid growth in the availability of multimedia content in digital form. They can be easily duplicated and redistributed. Thus there has been an increasing interest in developing copy protection or copy deterrence mechanisms. Digital watermarking has been proposed for the purpose of copyright protection for multimedia content. Digital watermark are some marks carrying the owners' information or the buyers' information, and they are embedded into the digital content without affecting the original copy's quality. Digital fingerprinting has been put forward in [Wag83]. The owner of the digital content embeds a unique digital watermark (called digital fingerprint) to each copy he sold. Thus, when the owner finds a redistributed copy, he can extract

*This paper is supported by National Key Fundamental Research Development Project(973, No.G1999035805), National High-tech Research Development Plan Youth Foundation (No.2001AA140447) and State Key Laboratory of Information Security Innovation Foundation.

the information embedded in the redistributed copy and the information will help him trace and identify the traitor.

However, the conventional fingerprinting schemes ([BMP85], [BS95]) are in symmetric sense, i.e., both the merchant and the buyer know the copy with the buyer's fingerprint. Thus a malicious merchant could spread himself the copy sold to some buyer and then accuse that buyer of having done so. And on the other hand, a buyer whose fingerprint has been found in the redistributed copy could claim that it is the owner who wants to frame her that redistributed the copy with her fingerprint deliberately. In order to solve this problem, Pfitzmann and Schunter proposed the concept of asymmetric fingerprinting in [PS96]. The above problem is solved in the following way: After a sale, only the buyer knows the copy with the fingerprint. However, if the merchant later finds this fingerprinted copy somewhere, he could identify the buyer and obtain a proof that the buyer is a traitor. This gives both the merchant and the buyer security in disputes.

Since the concept of asymmetric fingerprinting has been put forward in [PS96], quite a few papers have discussed the constructions of asymmetric fingerprinting schemes ([PW96],[PW97_a],[PW97_b],[BM97],[Domi99] and [Cam2000]). However, those constructions rely on either secure multi-party computations ([CDG87]) or zero-knowledge proofs(ref. chap4 of [G2001]) or other cumbersome techniques which are very inefficient to realize(we will address this in section 2 again). In this paper, we present a new asymmetric fingerprinting framework based on secret sharing under some assumptions about the underlying fingerprinting scheme. Our framework is rather efficient comparing to the previous constructions.

The structure of this paper is organized as follows: Section 2 describes some related work. In section 3, we put forward an asymmetric fingerprinting framework under some assumptions about the underlying fingerprinting scheme. Some comparisons and efficiency analysis are discussed in section 4. In section 5, we describe a fingerprinting scheme which could be used in the framework. Some further research directions are discussed in section 6.

2.   RELATED WORK

In 1996 Pfitzmann put forward the concept of asymmetric fingerprinting and presented a framework in which secure multi-party computation (SMPC) is used. It is a clever idea, but not an efficient way. In [PW96], [PW97_a] and [BM97], based on the coding of [BS95], they put forward asymmetric fingerprinting schemes which can resist lager collusions. However, in their methods, secure 2-party computations are still be used (though the authors put forward some suggestions, they are still not efficient in practical use). In [MW98], Memon and Wong construct an asymmetric fingerprinting protocol which avoids using SMPC, but their methods only applied to some limited cases where the watermark embedding methods are some operations like $\oplus$. Moreover, using their methods, one has to calculate the total digital content to be fingerprinted using the public-key encryption and decryption operations which are known to be much less efficient than operations in symmetric cryptosystems. When the digital content is large, their method will be quite inefficient. [PW97_b] uses the similar idea of [PW96] to get asymmetric property and the fingerprinting protocol also relies on secure 2-party computation. In [Domi99], Domingo has proposed an asymmetric fingerprinting scheme based on oblivious transfer protocol. But the oblivious transfer of the image blocks is needed, which is less efficient. What's more, in the identification protocol, all buyers may have to be asked to take part in the protocol, which is a burden and unnecessary for the innocent buyers. In [Cam2000], J.Camenisch presented an anonymous fingerprinting scheme based on group signature and as a by-product, he achieved an asymmetric fingerprinting scheme with 2-party trial. However, his asymmetric scheme still based on [PW96]. In this paper, we propose a new asymmetric fingerprinting framework. It avoids using SMPC and the zero-knowledge proof primitives, which are preferred by some previous asymmetric fingerprinting protocols. The basic idea of our framework is to use secret sharing when getting the keys to decrypt the blocks of the digital content. Those blocks are pre-calculated by the merchant: Some information is embedded in the blocks and they are sent out in encrypted form. When the users get the decrypted blocks, they get their corresponding codewords (we use the basic distribution scheme in [BS95]).

3. OUR FRAMEWORK FOR ASYMMETRIC FINGERPRINTING

3.1 Basic idea

In asymmetric fingerprinting protocols, the merchant doesn't know the exact fingerprinted copy the buyer purchases. That means there must be some information the merchant doesn't know. Only if he finds the redistributed copy, can he get enough proof to prove some buyer is a traitor. The traditional way to realize asymmetric fingerprinting protocols is to use secure 2-party computations. The buyer and the merchant secretly enter their secret information and after the secure computation, the buyer gets the copy he wants, and the merchant gets his record on the purchase. We note that, though there may be some implementation of SMPC in this context, non of them is quite efficient. So we devise an asymmetric fingerprinting protocol without using SMPC. In our framework, we assume the codewords assigned to the buyers are binary. (But in a very similar way, it can be extended to the case of non-binary.) First, the merchant cuts his original copy into l blocks (as in [BS95]), l is the length of the codeword. By embedding '0' and '1' respectively into $block_i$, he gets two versions: $block_i^0$, $block_i^1$ respectively. Then he uses two independent keys (keys in symmetric cryptosystem) to encrypt the two blocks respectively. He did the same thing for every block and all the keys are chosen independently. Then he sends out the $2l$ encrypted blocks. When the buyer decrypts $block_i^0$, the corresponding bit of her fingerprint is '0' and when decrypting $block_i^1$, the corresponding bit of her fingerprint is '1'. The basic idea of our framework is to let the merchant and the buyer share the key. Without help of the merchant, the buyer can't decrypt the blocks. And when the buyer asks the merchant for the key, the latter doesn't know the corresponding information of the buyer's codeword.

3.2   Our framework

Basic components:

The protocol includes 4 parties: the merchant (M), the buyer (B), the fingerprint issuing center (FIC) and the arbiter. FIC issues random binary codeword to the buyer and also issues the corresponding sub-keys of the secret sharing scheme to the buyer.

Cryptographic primitives to be used: a signature scheme and a bit commitment scheme. In our scheme, we assume that the fingerprinting scheme has reasonable collusive tolerance and tracing only relies on the merchant's initial information and his chosen random bits. i.e., without the help of buyers and without information on buyers' codeword, the merchant can deduce a possible traitor from the illegal copy and the information he stores.

There are 4 protocols: initialization, fingerprinting, identification and trial protocols.

I. *Initialization:* All parties generate their certificated public key and secret key pairs (Under assumed Certificated Authority) and publicize their corresponding public keys.

The merchant's pre-calculations: The merchant cuts his original copy into l blocks, l is the length of the codeword. By embedding '0' and '1' respectively into $block_i$, he gets two blocks: $block_i^0$, $block_i^1$. Then he uses two independent keys key_i^0, key_i^1 to encrypt the two blocks respectively. He did the same thing for every block and all the keys are chosen independently. He sends out the $2l$ encrypted blocks. The keys used is formed as follows:

For block$_i$, the two independent keys key_i^0, key_i^1 are selected from F_2^r, which are used as the keys for symmetric cryptosystem, e.g. DES,IDEA and AES. Then the merchant chooses another element $key_i^m \in F_2^r$ and computes $key_i^{f0} = key_i^0 \oplus key_i^m$, $key_i^{f1} = key_i^1 \oplus key_i^m$.

The merchant stores key_i^m and sends key_i^{f0} and key_i^{f1} to FIC as subkeys.

II. *Fingerprinting:*

1. FIC randomly selects a codeword $W_B \in F_2^l$ for the buyer B, signs using its secret key and then sends the codeword and signature to B .

2. B verifies the signature using FIC's public key and decrypts the codeword using her secret key. Then B commits to W_B bit by bit using bit commitment scheme and signs the commitments Com_B using her private key. The signature is denoted by Sig_B. B sends (Com_B, Sig_B) to M.

3. Using B's public key pk_B, M justifies Sig_B is a valid signature on Com_B by B. Then M sends Com_B to FIC for its correctness. FIC asks the buyer B to open Com_B and check whether the result equals to W_B and if this holds, protocol continues, otherwise protocol aborts. (Note we assume FIC executes this step honestly.) And for efficiency, FIC can randomly asks the buyer to open some bits in her commitment and check whether they equal to the corresponding bits in W_B.

4. FIC sends the sub-keys to B according to W_B. When the i-th bit of W_B is 0, FIC sends key_i^{f0}, otherwise key_i^{f1} ($i = 1$ to l).

5. B asks M to give her another part of the keys. i.e., M encrypts key_i^m ($i = 1$ to l) using B's public key pk_B and sends them to B.

6. B first decrypts the keys using her private key. Upon getting key_i^m, B can compute $key_i^0 = key_i^{f0} \oplus key_i^m$ or $key_i^1 = key_i^{f1} \oplus key_i^m$. Then she could get the corresponding decrypted blocks of the digital copy, in which the embedded information coincides with her codeword.

7. M stores (pk_B, Com_B, Sig_B) and other information to be used in traitor tracing.

III. *identification:*

When M finds a redistributed copy, using the information he stored, M identifies some possible traitors and some corresponding bits which will constitute the proof. For example, he identifies B and some possible bits (to say t bits) of her codeword, denoted by $Witness\_bits_B$ and their corresponding locations $Witness\_locations_B$ in the codeword. Those bits are called witness bits of B.

(Note that we have assumed using the underlying fingerprinting scheme, the merchant can deduce the possible traitor and some corresponding bits from the redistributed copy and the information he stores.)

IV. *trial:*

1. M sets
proof=$(pk_B, Com_B, Sig_B, Witness\_bits_B, Witness\_locations_B)$
and sends the proof to the arbiter.

2. The arbiter first justifies Sig_B is a valid signature on Com_B using B's public key pk_B. Then he asks the buyer B whose public key is pk_B to open the corresponding bit in Com_B and checks whether the corresponding bits equal to the bits in $Witness\_bits_B$. If all the bits pairs are equal, he judges the buyer B is illegal. Otherwise he judges M wants to frame B.

3.3   Security analysis

By means of security of an asymmetric fingerprinting scheme, we take the notions in [PS96]. It should be considered from three aspects: First, buyers could obtain useful data as long as nobody cheats. Second, if an illegal copy turns up, the merchant could identify some buyer who should be responsible for it. Third, buyers could be protected from cheating merchant and other buyers. We'd like to point out in practical environment, the third point is more important than the second. Because once an innocent buyer is accused of guilty, few people are willing to buy that merchant's products.

For our scheme, the first point is easy to satisfy if all the parties act according to the protocol. For the second point, we assume that under reasonable collusion size, using the underlying fingerprinting methods, the merchant could identify some traitor and several witness bits upon finding an illegal copy. The possibility the merchant could identify the traitor relies heavily on the underlying fingerprinting scheme.

For the third, we have to consider the security of the bit commitment scheme. A Bit Commitment(BC) scheme allows Alice to send something to Bob that commits her to a bit b of her choice in such a way that Bob can't tell what b is, but such that Alice can later prove him what b originally was. The commitment obtained after the commit phase is binding if Alice cannot change the bit and it is concealing if Bob cannot obtain any information about b without the help of Alice. The commitment is secure if it is binding and concealing. Here we assume the BC scheme is computationally secure. For several cryptographic realizations, see [BCC88].

We also note that in our asymmetric framework, the commitments have to be checked by FIC to prevent the buyer from concealing bit different from the corresponding bit in her codeword. Otherwise, the buyer may conceal different bits in order to deny her guilty in the trial protocol. Because the codewords assigned to the buyers are chosen randomly, thus the probability an innocent user will be framed is very small. We have the following proposition:

Proposition 1 If the codewords assigned to the buyers are chosen randomly and all the cryptographic primitives in use are secure, under the assumption that FIC executes the protocol honestly, then the probability an innocent buyer will be judged as a traitor is $1/2^t$.

Proof We take the proof under some computational assumptions (under which the cryptographic primitives we used are secure). If the merchant can frame an innocent buyer, he must get at least t witness bits of that buyer. If all the parties execute the protocol honestly, it means that he can get the corresponding bits in the commitment or he guess the correct bits. But the former is contrary to the security of the bit commitment scheme, by which M gets no information about the buyer's codeword. (Note that M also couldn't change the commitments of his will, because the buyer signs it and the arbiter could verify the signature.) So only M is able to guess at least t witness bits correctly, can he frame an innocent buyer. This probability will not exceed $1/2^t$. And if the underlying scheme makes t be large, the probability can be made negligible.

Note 1. In our protocol, we assume there are no active involvement of FIC, it can just act according to the protocol. In fact, the issuing of codewords, the issuing of sub-keys and the verification of W_B could be programmed into a tamper-resistant hardware which can run automatically when FIC supplies it with appropriate inputs, e.g., the users' codewords. For FIC, it is of little use to take time to break the hardware. Because even if he gets access to all the sub-keys, he can't get another part of the keys to decrypt the encrypted blocks. Comparing to the case that in some framework where the merchant has to supply his original copy to certain trusted third party, our framework ensures that FIC himself can't get the copy, because he only gets sub-keys. So our framework is more feasible in practical use.

4.   EFFICIENCY ANALYSIS

As we have mentioned, this framework is more practical than those schemes using SMPC. In [PS96], the author also made a realization for the asymmetric fingerprinting. But the method needs the commitments to all the content of the digital copy. So the computation is prohibitive for realistic data. Other comparisons can also be found in the related work section. Here we'll show the computations and storage requirements in our framework.

Computations: Except for the computations in embedding and extracting, the merchant will do symmetric encryptions for all the sub-blocks two times(because there are 2 versions for each block). But for each item to be sold, the merchant need only do these operations once for all buyers. Not like the methods in [PS96] and [Domi99], whenever a buyer buys a copy, the merchant has to do the fingerprint embedding for the buyer. So our method is of much value when the copy to be fingerprinted is large, which is usually the case in practical use. Buyers need only do XOR operations and decryptions of ciphertext in symmetric cryptosystem. As to the bit commitment, the merchant, buyers and FIC need only to take some computations for the bits in the codeword, which is rather efficient comparing to making commitments to all the digital content.

Storage scale: The encrypted blocks could be publicized on the internet or to be stored by M or FIC. M has to store sub-keys for each blocks of digital content and the selling records. FIC should store two times length of the sub-keys M stores. Buyers only needs to store keys temporarily and he also has to store some information used to open the commitments. E.g., if an item to be sold is divided into 512 segments and the key length is 128 bits, then the keys to be stored by M are 64K bits. So the store scale is very reasonable.

From the above analysis, our method is advantageous when the merchant has several valuable but rather large items to sell(for example, movies, DVDs). Because each segment is encrypted and decrypted using operations in symmetric cryptosystem, our methods can achieve more efficiency than those which need public-key cryptographic operations to the items or using other cryptographic primitives(e.g. BC). But when there are many little items to be fingerprinted, we don't recommend our method, because the storage scale may be too large if we use different keys for different items.

5. AN ASYMMETRIC FINGERPRINTING SCHEME BASED ON REPEATED CODING

In order to let our framework be more specific, we take a try to design a fingerprinting scheme in which M could identify a traitor without the help of other parties. The idea is to use a unique pseudo sequence to control the repeated embedding of each bits in a buyer's codeword. The idea may be somewhat similar to that in [CKLS97]. But our purpose is to design a simple coding method to help the merchant identify traitors and possible witness bits.

Here we also take the same assumption as in [BS95](the Marking Assumption): Colluding pirates can detect a specific mark location if and only if the mark differs between their copies. They cannot change an undetected mark without destroying the fingerprinted object.

The naïve idea of enabling the merchant to identify one witness bit is to use repeated coding, i.e., to embed one bit many times to ensure it can be correctly retrieved. But by simple repeated embedding, if the collusive buyers have 2 versions in one mark location, they have 2 versions in all the locations where the accompanying repetition bits of that mark lie in. So they can easily change half of them to their inverse, thus the bit is hard to retrieve. Furthermore, as we have shown above, the framework requires that tracing only relies on the merchant's initial information and his chosen random bits. But in the naïve idea, if the merchant doesn't know the corresponding original bits, it is rather hard to identify the correct bits. So the naïve idea of simple repeated embedding is of little use. We use the idea that the merchant uses pseudo random sequences (called control sequence and denoted by $control_B$ for buyer B) to decide half of the repetition bits will be embedded inversely. Suppose we embed one bit d times, then there will be about $d/2$ '0' and $d/2$ '1' and they distributed evenly in the resulting repeatedly embedded bits for that particular mark. So even if the collusive buyers' mark is different in one particular location in their codewords, they can't get this information by comparing their d repetition bits for the particular bit. For example(just to show the idea), if $d = 10$, Two collusive buyers B_1, B_2 have '0' and '1' respectively in one mark. If we simply embed them d times, B_1 and B_2 can easily detect the difference. But if we use the control sequence '11010 01011' (for B_1) and '01100 11010' (for B_2) respectively and the '1' in the control sequence indicates the corresponding bit of the original sequence will be reversed, the embedded information will be '11010 01011' (for B_1) and '10011 00101'(for B_2) respectively. By comparing the blocks, it is hard for them to decide whether the particular bit is the same. Furthermore, we can assume the embedding technique and the use of pseudo random sequence to control the inversion are kept secret from buyers. By the randomness of the control sequence, the buyer can only make some random choice in the detectable bits of the repetition bits, and by the Marking Assumption, they can't change the undetectable bits. So, for a colluder B, when using $control_B$ in the retrieval process to get back the original value of the repetition bits of the particular mark, the merchant will find more '0' or more '1', depending on B's mark in that particular location. However, for an innocent buyer B', when using $control_{B'}$ in the retrieval process, the merchant would see about half of the repetition bits of the particular mark are '0', and the rest are '1'. Thus, using the difference between '0' and '1', not only can the merchant differentiate a colluder from an innocent buyer, but also decide the original mark of the colluder with high probability by taking reasonable threshold value.

Most parts of the protocol using the above coding method are the same with those in the framework in section 3.2. But in order to let the merchant embed the mark inversely, the merchant will make pre-calculations for $L = ld$ blocks, where d is the repetition times of a single bit. The difference is: For block$_i$, (i is from 1 to L,) the merchant selects four independent keys $key_i^0, key_i^1, key\_r_i^0, key\_r_i^1 \in F_2^r$ and he chooses another two element $key_i^m, key\_r_i^m \in F_2^r$ and computes $key_i^{f0} = key_i^0 \oplus key_i^m$, $key_i^{f1} = key_i^1 \oplus key_i^m$ and $key\_r_i^{f0} = key\_r_i^0 \oplus key\_r_i^m$, $key\_r_i^{f1} = key\_r_i^1 \oplus key\_r_i^m$. $key_i^0, key\_r_i^1$ are used to encrypted block$_i^0$, $key_i^1, key\_r_i^0$ are used to encrypted block$_i^1$. (Note that $key\_r_i^0, key\_r_i^1$ and $key\_r_i^m$ are used for inversely embedding.) For one block, M gets 4 encrypted blocks. So M has to pre-calculate $4L$ encrypted blocks.

M stores key_i^m and $key\_r_i^m$ and sends $KEY0 = \{key_i^0, key\_r_i^0\}$ and $KEY1 = \{key_i^1, key\_r_i^1\}$ to FIC as sub-keys. In the fingerprinting protocol, FIC will send the sub-keys to B according to $W_B' = W_{B,1}W_{B,1}...W_{B,1}$ $W_{B,2}W_{B,2}...W_{B,2}......W_{B,l}W_{B,l}...W_{B,l}$ (every $W_{B,j}$ appears d times, $|W_B'| = dl = L$). When the i-th bit of W_B', ($i = 1$ to L) is 0, FIC gives $KEY0$, otherwise gives $KEY1$. M will send his sub-keys according the pseudo sequence for B. If the corresponding bits is 0, he sends key_i^m otherwise $key\_r_i^m$. Thus he can implement the reversely embedding of some bits while not knowing any information on B's codeword.

6.　　CONCLUSIONS

In this paper we present a new idea for asymmetric fingerprinting based on secret sharing. We also describe an idea for the underlying fingerprinting scheme. When the digital data to be fingerprinted is large, our framework is quite efficient. Our future research will include to devise fingerprinting methods satisfying the assumption for the underlying fingerprinting scheme in the framework. In addition, though the trust requirement for FIC is lower than that for a real TTP, it is still necessary to make some trust assumptions on FIC in our framework. Our future research will pay more considerations on the role of FIC and how to decrease the trust requirement on FIC. What's more, deep investigations on other secret sharing scheme like verifiable secret sharing may give us good suggestions for the constructions of asymmetric fingerprinting protocol.

References

[BCC88] G. Brassard, D. Chaum and C. Crepeau, Minimum disclosure proofs of knowledge,Journal of Computer and System Sciences(JCSS), 37(2), pp.156-189,1988.

[BMP85] G. R. Blakley, C. Meadows and G. B. Purdy, Fingerprinting Long For-giving Messages, Proceedings of Crypto'85, Springer-Verlag, pp.180-189, 1985.

[BM97] I. Biehl and B. Meyer,Protocols for Collusion-Secure Asymmetric Fingerprinting, Proc. 14th Annual Symposium on Theoretical Aspect of Computer Science, Springer-Verlag, pp.399-412,1997.

[BS95] D. Boneh and J. Shaw, Collusion-Secure Fingerprinting for Digital Data, Advances in Cryptology: Proceedings of Crypto'95, Springer-Verlag, pp.452-465, 1995.

[BS98] D. Boneh and J.Shaw, Collusion-Secure Fingerprinting for Digital Data, IEEE Trans.Inform.Theory, vol IT-44, pp.1897-1905, Sep.1998.

[CAM2000] J. Camenisch, Efficient Anonymous Fingerprinting with Group Signatures, In *Advances in Cryptology -Asiacrypt 2000*. volume 1976 of LNCS, Springer Verlag, pp.415-428, 2000.

[CDG87] David Chaum, Ivan B. Damgard and Jeroen van de Graaf, Multiparty Computations Ensuring Privacy of Each Party's Input and Correctness of the Result, Advances in Cryptology - CRYPTO'87 Proceedings, Springer-Verlag, pp.87-119, 1988.

[CKLS97] I.Cox, J.Killian, T.Leighton and T.Shamoon, Secure spread spectrum watermarking for multimedia, IEEE Transactions on Image Processing, vol. 6, no.12, pp.1673-1687, December 1997.

[Domi99] J. Domingo-Ferrer,Anonymous fingerprinting based on committed oblivious transfer,in Public Key Cryptography'99 (Lecture Notes in Computer Science 1560), eds. H. Imai and Y. Zheng, Berlin: Springer-Verlag, pp.43-52, 1999.

[G2001] O.Goldreich,Foundations of Cryptography:Volume 1-Basic tools, Cambridge University Press, 2001

[MW98] Nasir Memon and Ping Wah Wong,A Buyer-Seller Watermarking Protocol, IEEE Signal Processing Society 1998 Workshop on Multimedia Signal Processing, Los Angeles, California, USA. Electronic Proceedings, 1998.

[PS96] B. Pfitzmann and M. Schunter,Asymmetric Fingerprinting, Advances in Cryptology – EUROCRYPT'96, Springer-Verlag, pp.85-95, 1996.

[PW96] B. Pfitzmann and M. Waidner, Asymmetric Fingerprinting for Larger Collusions, IBM Research Report RZ 2857(# 90805)08/19/96, IBM Research Division, Zurich, 1996.

[PW97_a] B. Pfitzmann and M. Waidner, Asymmetric Fingerprinting for Larger Collusions, 4th ACM Conference on Computer and Communications Security, Zürich, pp.151-160, 1997.

[PW97_b] B. Pfitzmann and M. Waidner, Anonymous Fingerprinting, Advances in Cryptology – EUROCRYPT'97, Springer-Verlag, pp.88-102, 1997.

[Wag83] N. Wagner,Fingerprinting,Proceedings of the 1983 Symposium on Security and Privacy, pp.18-22, 1983.

AUTHENTICATION OF TRANSIT FLOWS AND K-SIBLINGS ONE-TIME SIGNATURE

Mohamed Al-Ibrahim
Center for Computer Security Research
University of Wollongong
Wollongong , NSW 2522, Australia
ibrahim@ieee.org

Josef Pieprzyk
Center for Advanced Computing - Algorithms and Cryptography
Computing Department
Macquaire University
Sydney, NSW 2109, Australia
josef@ics.mq.edu.au

Abstract   We exploit the unique features of the k-sibling Intractable hashing method in presenting two authentication schemes. In the first scheme, we propose a method which enables intermediate nodes in IP communication networks to verify the authenticity of transit flows. While in the second scheme, we introduce a new one-time digital signatures scheme.

Keywords: Network security, k-sibling hashing, source authentication, one-time signatures

1.   INTRODUCTION

There has been considerable interest in group-based applications over the last few years with the emergence of new sorts of communication modes such as multicast, concast and broadcast. There has also been a remarkable increase in real-time applications such as online video/audio streaming which have special quality of service requirements. As far as the security of these applications is concerned, new challenges in designing security protocols for these applications have arisen. Usually, these applications have special quality-of-service (QoS) requirements and the security services should be performed within its limits. One of the important security services is source authentication. Typical authentication schemes such as digital signatures have both high computational and space overhead, and hence they do not fulfill the new requirements of these applications. On the other hand, Message Authentication Codes (MAC) are more efficient, but does not provide non-repudiation service. Therefore, new techniques are required which not only can guarantee secure communication, but also maintains the efficiency of the application. This problem has been well defined and explored in the literature and several techniques have been proposed [2, 1, 3, 5, 14].

In this paper we continue the work in the direction of improving and developing efficient cryptography solutions for the problem of authentication in network communication. We introduce new authentication schemes that are based on the idea of the k-sibling intractable function family SIFF [8]. SIFF is a generalization of the universal one-way function family theorem (see also [15]). It has the property that given a set of initial strings colliding with one another, it is infeasible to find another string that would collide with the initial strings. This cryptographic concept has many useful applications in the security and we have exploit it to develop new authentication scheme. In this work, we start by expanding the idea of SIFF to hierarchical SIFF. Then, we proposed a scheme for authenticating 'transit' flows in IP communication. To our best knowledge, this topic has not been discussed elsewhere. Further, we propose a new one-time signature scheme that is efficient in generation and verification of signatures and with minimum space overhead, which is suitable for end-to-end real-time applications.

The paper is structured as follows. In the next section, we first illustrate the idea of k-SIFF and then expand it into a Hierarchical k-SIFF. In section 3, a scheme for authenticating transit flow in communication networks is illustrated. In section 4, the k-sibling one-time signature is presented.

2.   K-SIBLING INTRACTABLE HASHING

The construction and security properties of k-sibling intractable hash functions are discussed in [8]. Briefly, let $U = \cup_n U_n$ be a family of functions mapping $l(n)$ bit into $m(n)$ bit output strings. For two strings $x, y \in \sum^{l(n)}$ were $x \neq y$, we say that x and y collide with each other under $u \in U_n$, or x and y are siblings under $u \in \mathcal{U}_n$, if $u(x) = u(y)$.

in other words, sibling intractable hashing provides a hashing that collides for k messages selected by the designer. It can be seen as the concatenation of two functions: universal hash function and collision resistant hash function. More formally, we say that a family of universal hash functions

$$\mathcal{U} = \{U_n : n = \mathcal{N}\}$$

holds k-collision accessibility property if for a collection $\mathcal{X} = \{x_1, \ldots, x_k\}$ of k random input values $U_n(x_1) = \ldots = U_n(x_k)$ where $U_n : \{0,1\}^{\ell(n)} \to \{0,1\}^{L(n)}$ and $\ell(n), L(n)$ are two polynomials in n (n is the security parameter and $\mathcal{N}$ is the set of all natural numbers). A family of collision resistant hash functions

$$\mathcal{H} = \{H_n : n = \mathcal{N}\}$$

consists of functions that are one-way, and finding any pair of colliding messages is computationally intractable. k-sibling intractable hash functions can be constructed as

$$kH_n = \{h \circ u : h \in H_n, u \in U_n\}$$

where $U_n : \{0,1\}^{\ell(n)} \to \{0,1\}^{L(n)}$ and $H_n : \{0,1\}^* \to \{0,1\}^{L(n)}$ (the notation $\{0,1\}^*$ stands for strings of arbitrary length). U_n is a collection of polynomials over $GF(2^{\ell(n)})$ of degree k.

2.1 Hashing with a Single Polynomial

The designer of a k-sibling intractable hash function first takes an instance of a collision intractable hash $H : \{0,1\}^* \to \{0,1\}^\ell$ (such as SHA1) and a collection of k messages $\{m_1, \ldots, m_k\}$ that are to collide. Next she computes

$$x_i = H(m_i)$$

randomly chooses $\alpha \in GF(2^\ell)$ and determines a polynomial $U : \{0,1\}^\ell \to \{0,1\}^\ell$ such that

$$U(x_i) = \alpha \text{ for } i = 1, 2, \ldots, k.$$

This can be done using the Lagrange interpolation. Having k points $(x_i, \alpha); i = 1, \ldots, k$, it is easy to determine such a polynomial $U(x)$ and different from a straight line (see Appendix A). Note that, $k+1$ points are needed to determine a polynomial of degree k.

Denote $\mathbf{H} = U \circ H$. By construction $\mathbf{H}(m_i) = \alpha$ for all $i = 1, 2, \ldots, k$. The hash function $\mathbf{H}$ can be characterized by the following properties:

- finding collisions (those incorporated by the designer in U as well as those existing in H) is computationally intractable assuming the attacker knows descriptions of two functions H and U. The descriptions must be available in a public, read-only registry. Note that the description of U takes $k+1$ values from $GF(2^\ell)$.

- the hash function treats messages as an unordered collection of elements. To introduce an order, the designer needs to calculate $\mathbf{H}$ for an ordered sequence of messages so any message $m_i = (i, m_i')$ where i indicates the position of the message in the sequence. In other words, $\mathbf{H}(i, m_i') = \alpha$ for all $i = 1, \ldots, k$,

Note that if the number of colliding messages is large (say $k > 1000$), then to compute hash values, one would need to fetch $k + 1$ coefficients of polynomials $U(x)$. This introduces delays. Is there any other way to design k-sibling hashing?

2.2   Hierarchical Sibling Intractable Hashing

Given k-sibling intractable hash function $\mathbf{H}^{(k)}$ and a set $\mathcal{M} = (m_1, \ldots, m_{k^2})$ of messages. A k^2-sibling intractable hash function denoted as

$$\mathbf{H}^{(k^2)} = \mathbf{H}^{(k)} \circ \mathbf{H}^{(k)}$$

is a collection of $k + 1$ k-sibling intractable hash functions where

$$\mathbf{H}_i = U_i \circ H \text{ with collisions in } \mathcal{M}_i = (m_{ik+1}, \ldots, m_{(i+1)k})$$

for $i = 0, \ldots, k - 1$, and

$$\mathbf{H}_k = U_k \circ H \text{ with collisions in } \mathcal{X} = \{h_i = \mathbf{H}_i(\mathcal{M}_i); i = 1, \ldots, k\}.$$

To find hash value of a message, it is not necessary to know all polynomials U_i. For a message $m \in \mathcal{M}_i$, it is sufficient to know two polynomials only, namely, U_i and U_k.

In general, sibling intractable hashing with k^r colliding messages can be defined as

$$\mathbf{H}^{(k^r)} = \mathbf{H}^{(k)} \circ \mathbf{H}^{(k^{r-1})}$$

for $r > 2$. Similarly, to compute a hash value for a single message, it is necessary to learn r polynomials of degree k.

The polynomials $U_{i,j}(x)$ are in fact arranged in a tree structure. The leaves of the tree are $U_{1,j}$ for $j = 1, \ldots, k^{r-1}$. The next layer is created by polynomials $U_{2,j}$; $j = 1, \ldots, k^{r-2}$ and so on. The root is $U_{r,1}$.

Hierarchical sibling hashing holds the same security properties as the hashing with a single polynomial. The proof is relatively simple and follows the idea of Damgard's parallel hashing (see [7]).

3.   AUTHENTICATION OF PACKETS

Message authentication is an important service in information security. Typical authentication schemes such as digital signatures uses public-keys, while Message Authentication Codes (MAC), uses private keys. Digital signatures are known for their high computation overhead, while MAC does not provide non-repudiation service. In cases, such as in IP communication, we may have a stream of independent messages to be authenticated. Neither typical digital signatures provides efficient solution, nor MAC provides enough security service. Therefore, new techniques are required to provide both security and efficiency.

The other motivation is the requirement by the intermediate nodes in IP network for a technique to authenticate the packets in their transitions from source destination. IPSec [12] is a security mechanism designed to provide security services for IP communication. It provides source authentication, confidentiality, as well as integrity. As far as source authentication is concerned, with the symmetric authentication option provided by IPsec, only hop-by-hop authentication can be achieved. This means that a node that receives a message only knows that it came from an authenticated node where they share common key in the domain. However, it would not be able for intermediate nodes along the path to check the authenticity of the messages. In doing this, it would be possible to discover harmful actions such as denial of service attack in their early stages, before it is propagated to the destination. We seek a mechanism that enables intermediate nodes to verify the source of the message.

Possible solutions that may used for this case are for each message to be given a tag independent of one another, or for the concatenation of all messages to be given a single common tag. In the first method, the resulting tags may prove too impractical to be maintained, while in the second method the validation of one message requires the use of all other unrelated messages in recalculating the tag. A preferable method would be one that employs a single common tag for all the messages in a such a way that a message can be verified individually without involving other messages. This can be achieved by using SIFF, in which all messages are represented as a string of $l(n)$ bits long.

3.1   Authentication with Single Hashing

The security goal is to enable interested parties of the network (nodes) to authenticate messages (packets) in transit. A natural restriction imposed on authentication is that packets of the same message (generated by the same source) may travel through different routes. In effect, a node may see a small subset of all packets generated by the source. Those that are seen do not typically follow the initial order. Note that authentication of packets based on some sort of chaining is useless. Our solution is based on sibling intractable hashing.

Assumptions

- The source (sender) takes a message M and splits it into n packets (datagrams) In other words, $M = (m_1, \ldots, m_n)$ where m_i is the i-th datagram,

- There is a Public Key Infrastructure (PKI) that provides on demand authenticated public keys of all potential senders (normally in the form of certificates),

- The sender applies a secure signature scheme $SG = (S_{sk}, V_{pk}, G())$ for message authentication S_{sk} is the signing algorithm that for a given message m and a secret key sk produces a signature or $s = S_{sk}(m)$, V_{pk} is the verification algorithm that for a public key pk and a signature s returns 1 if the signature is valid and 0, otherwise. $G()$ generates a pair of keys: sk, pk. The meaning of "secure" will not be discussed here, and the interested reader can consult relevant papers (see [6]),

- The sender designs an instance of n-sibling intractable hash function $\mathbf{H}^{(n)}$ that is based on a collision intractable hash function H.

Sender

- Takes the sibling intractable hash $\mathbf{H}^{(n)}$ and computes the signature of the message M as

$$s = S_{sk}(H(H^{(n)}(M) \| H(u_0, \ldots, u_n)), \text{timestamp})$$

  where $U(x) = u_0 + u_1 x + \ldots + u_n x^n$ and $u_i \in GF(2^\ell)$,

- Puts the signature together with coefficients of $U(x)$ into a read-only registry R accessible to everybody.

Note that polynomial $U(x)$ must be used to produce the final hash value that is signed by the sender. This is done to prevent manipulation with the structure of the sibling intractable hash function $\mathbf{H}^{(n)}$.

Verifier

- Receives datagrams m_i where $i \in \{1, \ldots, n\}$,

- Contacts the registry R and fetches the signature and the coefficients u_i and recovers the polynomial $U(x)$,

- Obtains the authenticated public key pk of the sender from the PKI facility,

- Checks the validity of the signature using the algorithm $V_{pk}(s)$.

Note that the verification of the first datagram is the most expensive as it will take verification of signature (one exponentiation if signature is based on RSA or ElGamal) that comprises also the calculation of hash $h_i = H(m_i)$, computation of $U(h_i)$ and evaluation of $H(u_0, \ldots, u_n)$. Any new message can be verified using one extra evaluation of H and U. It computes the hash $h_i' = H(m_i)$ and computes $h' = U(h_i')$. If $h' = h$ then it accepts the message; otherwise, it rejects it. As far as communication is concerned, the verifier must fetch the signature and the polynomial $U(x)$. Note that the length of $U(x)$ is almost the same as the whole message M. This seems to be the weakest point of the construction.

3.2 Authentication with Hierarchical Hashing

In this case, the sibling intractable hashing is computed using a family of polynomials $U_{i,j}$ with $i = 1, \ldots, r$ and $j = 1, \ldots, k^{r-i}$. The message consists of $n = k^r$ datagrams. To compute $H^{(n)}(M)$ it is enough to fetch r polynomials of degree k (that is a sense, a path between a leaf and the root. If we choose $k=2$, then the verifier needs to fetch $3 \times \log_2 n$ coefficients. With pre-determined single points for the polynomials, the number can be reduced to $2 \times \log_2 n$ without security deterioration.

The tree of polynomials must also be subject to hashing (to make the verifier sure that she uses the correct instance of the sibling intractable hash). One good feature is that the verifier would like to use explicitly all polynomials she has imported from R. The signer may help the verifier by first using parallel hashing for the polynomials and storing in R all intermediate results of hashing. The verifier puts the polynomials together with intermediate hash values to generate $H(U)$ where U means collection of all polynomials.

The advantage of hierarchical hashing is evident when we consider the storage required to allow authentication of k public values using the following approach. An entity A authenticates t public values $Y_1, Y_2, \ldots, Y_t$ by registering each with a read-only registry or trusted third party. This approach requires registration of t public values, which may raise storage issues at the registry when t is large. In contrast, a Hierarchical Hashing requires only a single value registered in the registry. If a public key Y_i of an entity A is the value corresponding to a leaf in an authentication tree, and A wishes to provide B with information allowing B to verify the authenticity of Y_i, then A must (store and) provide to B both Y_i

and all hash values associated with the authentication path from Y_i to the root; in addition, B must have prior knowledge and trust in the authenticity of the root value R. These values collectively guarantee authenticity, analogous to the signature on the public-key certificate. The number of values each party must store is $\log(t)$.

Consider the length (or the number of edges in) the path from each leaf to the root in a binary tree. The length of the longest such path is minimized when the tree is balanced. i.e., when the tree is constructed such that all such paths differ in length by at most one. The length of the path from leaf to the root in a balanced binary tree containing t leaves is about $\log(t)$.

Using a balanced binary tree as authentication tree, with t public values as leaves, authenticating a public value therein may be achieved by hashing $\log(t)$ values along the path to the root.

Authentication trees require only a single value which is the root value, in a tree to be registered as authentic, but verification of the authenticity of any particular leaf value requires access to and hashing all values along the authentication path from leaf to root.

To change a public (leaf) value or add more values to an authentication tree requires re-computation of the label on the root vertex. For large balanced tree, this may involve a substantial computation. In all cases, re-establishing trust of all users in this new root value is necessary.

The computational cost involved in adding more values to a tree may motivate constructing the new tree as an unbalanced tree with the new leaf value being the right child of the root, and the old tree, the left. Another motivation for allowing unbalanced trees arises when some leaf values are referred far more frequently than others.

3.3   Security Issues

There follow some remarks on the security of the schemes:

- the scheme signs simultaneously all datagrams using a single signature. The important difference of this scheme from other schemes is that verification of datagrams can be done independently (or in parallel). In other words, to authenticate datagrams, one does not need to know all datagrams,

- no authentication is required for the coefficients fetched from read-only registry. This is because, if entries are tampered with, then packets will be rejected since the final hash recovered from the signature will be different from the hash value obtained from the datagrams and the polynomial,

- the only security problem could be of denial of service when an attacker may intentionally modify polynomial coefficients to reject the datagrams,

- in both flat and hierarchical k-sibling approaches, a single signature is required:

  1 the description of public polynomial coefficients used in the k-sibling intractable hashing takes about n integers each of size 160 bits for SHA1, where n is the number of packets of the message M and $k > 2$. If $k=2$ then this number $=2n$.

  2 the scheme may used against denial-of-service attacks. In particular, it would be able for receivers' at the intermediate nodes to ignore those packets that have failed to pass k-sibling hashing verification. attack from malicious source.

- the authentication scheme described above could be used for both types of IP data transfer modes: connection-oriented and connection-less. In the case of connection-oriented communication, where a node or destination sees almost all the packets of the message, flat sibling hash with a single polynomial $U(x)$ of degree n is best applicable. If, however, a node may see only a small fraction of packets, as in connection-less communication, then the hierarchical sibling with 2-sibling hashing seems to be superior.

4.   ONE-TIME SIGNATURES

One-time signatures derived their importance from their fast signature verification, in contrast to typical digital signature schemes, which have either high generation or verification computation time. One-time signatures are a perfect option for authenticating particular types of applications were receivers are of low power capability, such as smart cards, or for online applications, such as video/audio streaming, which requires fast verification.

One-time signatures have to be efficient and secure. Typically, the verification of the signature is expected to be very efficient. Additionally, signatures have to be initialized well ahead of time when messages are to be signed and verified. This allows the signer to pre-compute the signature parameters so they can be fetched by potential verifiers. Once the message is known, the signer can sign it quickly and receivers can verify the signed message efficiently. A distinct characteristic of one-time signatures is that they are used once only. To sign a new message, the signer must initialize the signature parameters (parameters of old signatures must not be reused). The security of one-time signatures is measured by the difficulty of forging signature by an adversary who normally has access to a single pair: message and its signature.

Rabin [11], Merkle [10] and GMR [13] are well known examples of one-time signature schemes. Although they differ in their approaches, but they share the same idea: only one message can be signed using the same key. Once the signature is released, its private key is not used again, otherwise, it is possible for an adversary to compute the key.

One of the new approaches in designing such signatures is BiBa one-time signature [5]. BiBa is an acronym for BIns and BAlls. It uses bins and balls analogy to create a signature. To sign a message m, the signer first uses random precomputed values generated in a way that a receiver can authenticate them with a public key. These precomputed values are called SEALS. (SElf Authenticating vaLues). The signer then compute the hash of the message $h = H(m)$) and then compute the hash function G_h. Now, the collision of SEALS under a hash function G_h forms a signature: $G_h(s_i)=G_h(s_j)$, where $s_i \neq s_j$. The BiBa signature exploits the birthday paradox property, in that the signer who has a large number of balls finds a collision (signature) with high probability, but a forger who only has a small number of balls has a negligible probability of finding a signature.

The BiBa signature scheme has desirable features such as small authentication space overhead and fast verification time. However, its public keys are very large, and the time needed to generate a signature is higher than any other known system, and it requires parallel processors to find collision of SEALS. This makes signature generation a computation overhead. Also, it uses an *ad-hoc* approach to find collisions among the 'SEALS' to the corresponding bin which results the high signature generation time.

4.1   *K*-Sibling One-time Signature

We propose a variant approach of BiBa by using the SIFF method. SIFF provides hashing with a controlled number of easy-to-find collisions. In other words, we apply a *deterministic* approach in finding a collision (signature).

As for signatures based on public-key cryptography, we assume that we are going to produce signatures for digests of messages. Thus suppose that messages to be signed are of constant length (160 bits if we use SHA1).

Let $\text{SIFF}_i(x)$ be an instance of k-sibling hash function that for k inputs $x_{i,0}, \ldots, x_{i,k-1}$ produces the output α_i or

$$\text{SIFF}_i(x_{i,j}) = \alpha_i \text{ for } j = 0, \ldots, k-1$$

The function applies a polynomial $U_i(x) = u_{i,0} + u_{i,1}x + \ldots + u_{i,k-1}x^{k-1}$ that collides for the inputs $x_{i,0}, \ldots, x_{i,k-1}$ or

$$U_i(H(x_j)) = \alpha_i$$

where H is a collision-resistant hash function. Assume that the message to be signed is $M = (m_1, \ldots, m_t)$ where m_i are v-bit sequences. The message M consists vt bits (typically of the length 160 bits).

To design our one-time signature we use the sequence of t instances of SIFF where each instance applies 2^v collisions.

Initialization
The signer builds up the sequence of $\text{SIFF}_i(x)$ for $i = 1, \ldots, t$. He starts from $\text{SIFF}_1(x)$. First he picks up at random a sequence of 2^v integers (whose length is determined by the security parameter of the signature). Let the sequence be $r_{1,j}$; $j = 0, \ldots, 2^v - 1$ and denote $x_{1,j} = (r_{1,j}, j)$. The signer chooses at random the output α_1 and calculates the polynomial $U_1(x)$ such that

$$U_1(H(x_{1,j})) = \alpha_1 \text{ for } j = 1, \ldots, 2^v$$

Next, he creates $\text{SIFF}_i(x)$ for $i = 2, \ldots, t$. For each i, he selects at random integers $(r_{i,j})$; $j = 0, \ldots, 2^v - 1$, composes

$$x_{i,j} = (r_{i,j}, j, \alpha_{i-1})$$

and calculates the polynomial $U_i(x)$ such that

$$U_i(H(x_{i,j})) = \alpha_i$$

for a random α_i. The polynomials $U_i(x)$ and the final value α_t are made public in the read-only authenticated registry; $i = 1, \ldots, t$.

Signing
Given a message $M = (m_1, \ldots, m_t)$. The signer marks the input x_{1,m_1} and extracts r_{1,m_1} and similarly determines r_{i,m_i} for $i = 2, \ldots, t$. The signature is

$$S(M) = (r_{1,m_1}, \ldots, r_{t,m_t})$$

The pair $(M, S(M))$ is the signed message.

Verification
The verifier takes the pair $(\tilde{M}, S(\tilde{M}))$ and the public information, i.e. coefficients of polynomials $U_i(x)$ and α_t. Knowing $\tilde{x}_{1,\tilde{m}_1} = (\tilde{r}_{1,\tilde{m}_1}, \tilde{m}_1)$ and the polynomial $U_1(x)$, he can compute $\tilde{\alpha}_1$. Next, he recreates the inputs $\tilde{x}_{i,\tilde{m}_i} = (\tilde{r}_{i,\tilde{m}_i}, \tilde{m}_i, \tilde{\alpha}_{i-1})$ for $i = 2, \ldots, t$. If the last recovered $\tilde{\alpha}_t$ is equal to α_t recovered from registry, the signature is considered valid. Otherwise, it is rejected.

Suppose that an adversary knows a signed message and tries to modify either message or signature such that the forged (and signed) message passes verification. Obviously, the adversary also knows the public information. Informally, if the adversary is successful it means he was able to create either a new collision (which was not designed by the signer) or was able to guess one of the strings $r_{i,m'}$. The first event is excluded if we assume that the SIFF is collision resistant. The probability of the second event can be made as small as requested by choosing an appropriate length of the strings $r_{i,j}$. It is important to note that the above considerations are valid only if the public information about signatures is authentic.

The signature allows to trade efficiency of verification with the workload necessary to set up the signature system. This is very important aspect of the signature. Note, however, that the setup is done for each single message (this is one-time signature). Verification is done many times, typically as many times as there are different recipients. Consider two extreme cases: the first with the longest signature where SIFFs are designed for binary messages or $v = 1$ (t is the largest) and the second with $t = 1$. The first case permits a very efficient setup of the system with relatively small public information. The price to pay is bandwidth necessary to transport a very long signature and verification consumes a largest number of hash operations (as many as bits in the signed message).

The second case applies a relative small number of SIFFs (t is small). The setup is very expensive as any single SIFF applies large number of collisions and in effect the corresponding polynomials are very long. Receivers must fetch polynomial coefficients for verification. Verification seems to be fast as it requires a small number of hash operations. Signatures are relatively short.

Some scope for more efficient implementation exists if the strings $r_{i,j}$ are generated differently. Note that the system applies $t2^v$ such strings but only t are used as the signature. To reduce the necessary storage for keeping the strings by the signer, it is reasonable to choose at random $t + 1$ integers $r_{i,1}$; $i = 1, \ldots, t$ and the integer R. A polynomial $G(x)$ of degree t can be design such that $G(0) = R$ and $G(i) = r_{i,1}$ for $i = 1, \ldots, t$. Note that other $r_{i,j}$ can be derived from the polynomial $G(x)$. This way of generation of $r_{i,j}$ is secure in the sense that signatures reveal t points on $G(x)$ leaving a single point on $G(x)$ unknown to the adversary.

5. CONCLUSION

The k-sibling intractable hashing function is a useful crytographic tool that might be used to solve a number of problems. We have exploited it to design a new authentication scheme that can verify the authenticity of independent messages. For example, it enables, intermediate nodes in a communication network to authenticate the source of packets. We have also used it to design a new one-time digital signature which has low computation and space overhead.

Appendix: A. Lagrange Interpolation Polynomial

The Lagrange interpolating polynomial is a polynomial of degree $n-1$ which passes through the n points $y_1 = f(x_1)$, $y_2 = f(x_2)$, $\ldots$, $y_n = f(x_n)$. It is given by:

$$P(x) = \sum_{j=1}^{n} P_j(x)$$

where,

$$P_j(x) = \prod_{k=1}^{n} \frac{x - x_k}{x_j - x_k} y_j$$

Written explicitly,

$$
\begin{aligned}
P_j(x) \quad = \quad & \frac{(x - x_2)(x - x_2)\ldots(x - x_n)}{(x_1 - x_2)(x_1 - x_3)\ldots(x_1 - x_n)} y_1 \\
+ \quad & \frac{(x - x_1)(x - x_3)\ldots(x - x_n)}{(x_2 - x_1)(x_2 - x_3)\ldots(x_2 - x_n)} y_2
\end{aligned}
$$

$$\vdots$$

$$+ \quad \frac{(x - x_1)(x - x_2)\dots(x - x_{n-1})}{(x_n - x_1)(x_n - x_2)\dots(x_n - x_{n-1})}y_n$$

References

[1] R. Gennaro and P. Rohatchi, " How to Sign Digital Streams", *Advances in Cryptology - CRYPTO'97*, Lecture Notes in Computer Science 1249, Springer-Verlag, pp. 180-197, 1997.

[2] R. Canetti, J. Garay, G. Itkins, D. Micciancio, M. Naor, B. Pinkas, " Multicast Security: A Taxonomy and some efficient Constructions", IEEE INFOCOM'99.

[3] A. Perrig, R. Canetti, J.D. Tygar, D. Song, " Efficient Authentication and Signing of Multicast Streams over Lossy Channels", ACM CCS'02 , 2000.

[4] M. Al-Ibrahim and J. Pieprzyk, "Authenticating Multicast Streams in Lossy Channels Using Threshold Techniques," in *Networking – ICN 2001, First International Conference,* Colmar, France, *Lecture Notes in Computer Science,* vol. 2094, P. Lorenz (ed), pp. 239–249, 2001.

[5] A. Perrig. "The BiBa One-Time Signature and Broadcast Authentication Protocol". ACM, CCS'01, Philadelphia, November 2001.

[6] M. Bellare, A. Desai, D. Pointcheval and Rogaway. "Relations among notions of security for public-key encryption schemes". In H. Krawczyk, editor, *Advances in Cryptology - CRYPTO'98*, Lecture Notes in Computer Science No. 1462, Springer-Verlag, pages 26-45, 1998.

[7] Ivan Damgard. "A design principle for hash functions." In G. Brassard, Editor, *Advances in Cryptology - CRYPTO'89,* Lecture Notes in Computer Science No. 435. pages 416-427. Springer-Verlag, 1989.

[8] Y. Zheng, T. Hardjono and J. Pieprzyk. "The sibling intractable function family(SIFF): notion, construction and applications. *IEICE Trans. Fundamentals,* vol. E76-A:4-13, January 1993.

[9] J. Pieprzyk and E. Okamato. " Verifiable secret sharing", *ICISC'99*, Seoul, Korea, pages 169-183, Lecture Notes in Computer Science No. 1787. Springer-Verlag 1999.

[10] R. Merkle. " A Certified Digital Signature", *Advances in Cryptology - CRYPTO'89,* pages 218-238 Springer-Verlag, 1989. Lecture Notes in Computer Science No. 435.

[11] A. Menezes, P. Oorschot and S. Vanstone. "Handbook of Applied Cryptography", CRC Press, 1996.

[12] S. Kent and R. Atkinson. RFC 2401: Security architecture for the Internet Security. Networking Group, IETF, November 1998. http://www.ietf.org.

[13] S. Goldwasser, S. Micali, and C. Rackoff, " A Digital signature Scheme Secure Aagainst Adaptive Chosen-message Attacks", SIAM Journal on Computing, 17 (1988).

[14] P. Rohatchi. " A compact and Fast Hybrid Signature Scheme for Multicast Packet Authentication", in *Proc. of 6th ACM conference on Computer and Communications Security,* 1999.

[15] M. Bellare, P. Rogaway. " Collsion-Resistant Hashing: Towards Making uniersal one way hash functions practical". *Advances in Cryptology - CRYPTO'97,*

Lecture Notes in Computer Science No. 1294, pages 470-484, Springer-Verlag, 1997.

[16] A. Shamir, "How to share a secret", *Communications of the ACM*, 22:612-613, November 1979.

[17] Oded Goldreich, Shafi Goldwasser, and Silvio Micali, " How to construct random functions", *Journal of the ACM*, 33(4):792-807, October 1986.

[18] S. Even, O Goldreich, S. Micali. "On-line/Off-line digital signatures", *Journal of Cryptology*, volume 9, number 1, winter 1996.

IMPROVING THE FUNCTIONALITY OF SYN COOKIES

André Zúquete
IST / INESC-ID Lisboa, Lisboa, Portugal
andre.zuquete@gsd.inesc-id.pt

Abstract   Current Linux kernels include a facility called TCP SYN cookies, conceived to face SYN flooding attacks. However, the current implementation of SYN cookies does not support the negotiation of TCP options, although some of them are relevant for throughput performance, such as large windows or selective acknowledgment. In this paper we present an improvement of the SYN cookie protocol, using all the current mechanisms for generating and validating cookies while allowing connections negotiated with SYN cookies to set up and use any TCP options. The key idea is to exploit a kind of TCP connection called "*simultaneous connection initiation*" in order to lead client hosts to send together TCP options and SYN cookies to a server being attacked.

Keywords: SYN flooding attacks, SYN cookies, TCP options, simultaneous connection initiation.

1.   INTRODUCTION

Current Linux kernels include a facility called TCP SYN cookies, designed and first implemented by D. J. Bernstein and Eric Schenk [1, 2]. This facility was conceived to face a specific attack on normal TCP/IP networking, known as "*SYN flooding attack*". This is a denial-of-service attack that floods a server host with long-lasting half-open TCP connections. Since kernels usually restrict the number of half-open connections, a SYN flooding attack prevents real client hosts from connecting to a server host being attacked. Furthermore, such attacks are easy to deploy, can be launched anywhere on the Internet and are difficult to avoid by well-known servers.

SYN cookies provide protection against this type of attack and their rationale is straightforward: prevent the denial-of-service scenario by not keeping state about pending connection requests, or half-open connections. The cookies allow a server host to maintain the state of half-open connections outside its memory: such state is (partially) stored inside a cryptographic challenge, the SYN cookie, that is returned to the client within SYN-ACK segments as the server's TCP Initial Sequence Number (ISN). Since TCP requires the client to send back that ISN on the subsequent ACK, the server will be able to restore a half-open connection from a cookie and, consequently, create a final connection descriptor.

Although the rationale behind SYN cookies is straightforward, its implementation is more complicated. First, cookies must fit in the space defined for the ISN field of a TCP header (32 bits). Second, the generation of cookies should respect the TCP recommendations for ISN values being monotonically increasing over time, possibly using some sort of time-based counter, to reduce the probability of delayed segments being accepted by new incarnations of similar connections [3]. Third, cookies must be unpredictable by attackers, to prevent the forgery of valid cookies, which could be used to launch TCP hijacking attacks [4]. And fourth, cookies should contain all TCP options sent by clients on SYN segments and usually kept in half-open connection's descriptors (if supported by servers).

The current implementation of the SYN Cookie Protocol (SynCP hereafter) in Linux systems deals differently with these issues, handling completely or partially some of them, but, unfortunately, ignoring most of them [1]. For instance, some TCP options that are relevant for throughput performance, like large windows or selective acknowledgment, are simply not supported when using SYN cookies. Even the choice of suitable server's Message Size (MSS) is limited. Therefore, a SYN flooding attack, by triggering the use of SYN cookies, can reduce the quality of service provided by a server host (besides forcing the waste of CPU cycles and bandwidth to deal with bogus connection requests). This is clearly an undesirable side effect of using SYN cookies.

In this paper we present an improvement of the current SynCP. This improvement uses most of the current mechanisms for generating and validating cookies; however, it allows connections negotiated with SYN cookies to set up and use TCP options that are relevant for performance but currently ignored. The key idea is to explore a kind of TCP connection called "*simultaneous connection initiation*". But this approach, although fully compatible with standard TCP rules [3], faces two major problems. First, some systems do not deal correctly with simultaneous connection initiations (e.g. Windows systems). Second, client-side firewalls may interfere with the action taken by the server. To overcome these problems we propose a mixed protocol, combining the current and the new SynCPs. Problematic clients are detected and handled differently by the server using a simple cache with their IP addresses.

This paper is organized as follows. Section 2 presents some related work regarding the countermeasures for SYN flooding attacks. Section 3 briefly describes how SYN cookies are currently handled by Linux kernels. Section 4 presents our proposal, starting with the basic protocol, presenting some problems it faces with problematic clients and firewalls, and concluding with the description of the final protocol. Section 5 sketches some implementation details. Finally, in Section 6 we draw some conclusions.

2. RELATED WORK

The generic goal of any solution to SYN flooding attacks is to continue to accept connection requests even when being under attack. There are several ways to achieve this goal, but none of them is perfect. In this section we will shortly describe the approach followed by several proposed solutions, presenting their advantages and drawbacks. The description will focus only on solutions that change the way server kernels deal with the current IPv4 TCP protocol specifications. We will not address any solutions requiring either (i) a modification of the TCP protocol, (ii) a modification of the kernel of client hosts or (iii) filtering policies applied to client hosts in their access to the Internet (e.g. [5]).

One obvious solution, proposed by several vendors, is to use larger queues for pending connection requests. A high bound for the queue's length can be computed from the bandwidth of the server's network connection and the timeout used by the server to discard pending requests. This is a sort of brute-force solution that may waste lots of kernel memory and slow down the server's response time, but it can be effective in public servers serving large communities of clients, since such hosts usually have extensive hardware resources.

A more crafty solution is called Random Drop [6]. The principle is simple: a server always accepts a new connection request and uses the queue of pending requests as a cache, with a random substitution policy to get space for new requests. For each dropped request the server sends an RST to the source host, enabling real clients to react to the server action. This solution allows a flexible trade-off of defense effectiveness with resource requirements, but it only guarantees service in a probabilistic manner. Thus, an attacker may still occasionally affect connections requested by real clients.

Another policy for dropping connection requests is called Reset Cookies [7, 6], using security associations[1] for improving correctness. When under attack, the server host checks a cache of security associations prior to accept and queue each new connection request. If the client is not listed in the cache the server replies with an illegal SYN-ACK with a cookie as the server's ISN. Legitimate clients will reply with an RST containing that cookie, which can then be checked by the server and trigger the creation of a security association with the client. This solution implies the storage and management of a cache of security associations in the server and increases significantly the latency of connections for clients not listed in the cache. Therefore, heavily used public servers should use large caches for reducing the impact of using Reset Cookies, but that is exactly the opposite of what defenses against SYN flooding attacks should do.

3.   SYN COOKIES IN CURRENT LINUX KERNELS

The most recent Linux kernel (version 2.4.14), as well as many other previous versions, allows kernel builders to include the generation and analysis of SYN cookies in the kernel functionality, and allows administrators to dynamically activate their use. We will now explain how SYN cookies are used on a kernel were they are enabled, in order to introduce all the problems currently raised by their use. For simplicity, hereafter we will use the term cookie to refer to a SYN cookie.

3.1.   Generation and validation of SYN cookies

Cookies are not used when the kernel operates in normal conditions, but only when it suspects of being under a SYN flooding attack[2]. The suspicion is simply derived from the length of queue of pending connection requests: if it reaches a given threshold length, the kernel emits a SYN flood warning and starts using cookies to handle new connection requests.

Cookies have a limited lifetime. When the kernel receives an ACK from a client (the third segment of the three-way handshake), first it checks the segment against the queue of connection requests, and upon failure it may check whether the segment carries a valid cookie. Such

---

[1]A security association is the IP address of a real client host that initiated a TCP connection to the server in the past.

[2]This is not true if the kernel operates as a so called *"SYN cookie firewall"*. In this case, all incoming SYN segments get a SYN cookie reply from the firewall kernel, and only upon the correct reception of the cookie within an ACK the server is contacted by the firewall.

checking takes place only within a short time frame, starting when the last cookie was sent and ending a few seconds later (currently 3 seconds), and cookies are accepted only if their value is among a set of acceptable values.

3.2.   Transparent use of SYN cookies

Cookies were designed to tackle SYN flooding attacks without changing TCP implementations used by client hosts; only servers using them must be modified in order to produce and validate them, and their use should be as transparent as possible to clients. Therefore, the use of cookies must conform with all TCP mandatory rules, but may impose restrictions on the use of TCP optional behaviours. These restrictions, however, should be minimized in order to reduce the side effects of using cookies.

Cookies are 32-bit values stored as ISN values in the sequence field of SYN-ACK segments sent by servers, and are retrieved from the sequence numbers acknowledged in ACK segments sent by clients. Therefore, cookies should respect the recommendations for the generation of ISN values [3, 4].

Cookies also carry some TCP options negotiated exclusively on SYN segments. Currently they only carry a 3-bit encoding of 8 predefined MSS values. All other options are simply ignored[3], including window scaling, selective acknowledgment, and time stamping for Round-Trip Time Measurement (RTTM) or Protection Against Wrapped Sequence numbers (PAWS). Unfortunately, all these options were introduced to improve the performance of TCP connections [8, 9]. Thus, we can conclude that a SYN flooding attack, by triggering the use of cookies, has the potential side effect of reducing the performance of some server's TCP connections.

3.3.   SYN cookies algorithms

The algorithm for generating cookies should try to reconcile two different goals. On one hand, cookies should be hard to guess by clients, in order to defeat attacks using ACK segments with forged, valid cookies. This implies that cookies should contain a large number of bits generated using servers secrets and functions not easily invertible by clients. On the other hand, cookies cannot be fully random and still respect the TCP rule of slowly growing over time. This implies that some part a

---

[3]SYN cookies cannot also handle correctly T/TCP connection requests. However, since this is still an experimental protocol, we will not address it further.

cookie should be generated without using values produced by crypto-graphic functions.

However, since cookies are only 32 bit long, it is difficult, if not impossible, to accomplish both goals simultaneously. Therefore, the algorithm to produce cookies cares only about security, and is completely independent of the algorithm to produce ordinary ISN values.

The algorithms currently used to generate and validate cookies are fully explained in Appendix A. In short, cookies are computed using constant secret values, TCP/IP addresses and ports of the client's SYN segment, a time counter and a 3-bit encoding of the server's MSS value (see Table A.1 in Appendix A). The validation of a cookie involves retrieving and testing the last two – time counter and MSS encoding – using the same secrets and the same fields of the client's ACK segment.

3.4.   Risk analysis

Cookies were devised to solve a problem, and not to create new ones. Thus, they should not allow attackers to launch other kinds of attacks using them. This means that attackers should not be able to produce valid cookies, since that would allow them to create fictitious TCP connections on a victim server.

The reality is that valid cookies are relatively hard to forge[4]. On a given instant, a valid cookie for a given pair of TCP addresses can only take 32 values out of 2^{32} possible ones. The value of 32 comes out of multiplying the 4 acceptable values for the time counter with the 8 possible values of the MSS encoding (see Appendix A for more details). Any increment in the range of either one of these values would naturally improve the probability of guessing valid cookies.

4.   OUR PROPOSAL

As previously mentioned, the generation of cookies is not a trivial task because one has to trade-off several different requirements. In this section we will show how the current SynCP can be improved in order to better deal with one of those requirements, namely the support of TCP options, without reducing its current functionality or its security against guessing attacks. Furthermore, we want to keep the basic approach of the current SynCP of not storing any state on servers, namely TCP options, for ongoing connection handshakes requested during a SYN flooding attack.

---

[4]Assuming that no better strategy exists for producing valid cookies besides random guessing.

The rationale for the new approach is the following: as TCP options cannot be fully embedded in cookies, for both practical and security reasons, then one has to force the client host to send again the TCP options together with the segment that carries the cookie. This means that the client must send another segment containing both the cookie and the SYN bit set, as only such segments may contain the TCP options that we are concerned with. This requirement can be met using the *"simultaneous connection initiation"* described in the seminal TCP documentation [3].

4.1. Basic approach

Figure 1 shows the diagram presented in [3] (and corrected in [10]) describing the steps followed in one simultaneous connection initiation. The new SynCP will explore this particular way of negotiation; in particular it will conduct the client socket through the same state transition of socket A of Figure 1.

	socket A state		segment		socket B state
1	CLOSED				CLOSED
2	SYN-SENT	→	(SEQ=x)(CTL=SYN)	...	
3	SYN-RCVD	←	(SEQ=y)(CTL=SYN)		SYN-SENT
4	...		(SEQ=x)(CTL=SYN)	→	SYN-RCVD
5		→	(SEQ=x)(ACK=y+1)(CTL=SYN,ACK)	...	
6	ESTABL.	←	(SEQ=y)(ACK=x+1)(CTL=SYN,ACK)	←	SYN-RCVD
7		...	(SEQ=x)(ACK=y+1)(CTL=SYN,ACK)	→	ESTABL.

Figure 1.   Steps followed by TCP sockets in a simultaneous connection initiation.

The new SynCP works as follows (see Figure 2-II). When the server receives a SYN, it computes a normal SYN-ACK reply, gets a cookie for the server's ISN, and sends it as a pure SYN (with the ACK bit disabled). A genuine client socket for the requested connection is in SYN-SENT state, will move to SYN-RCVD and reply to the server's SYN with a SYN-ACK, repeating its ISN number and all the TCP options sent by the server. When the server receives a SYN-ACK for a socket in LISTEN state it checks if the acknowledged sequence number is a valid cookie. If it is valid, the server creates a new connection with the client, and sends back a SYN-ACK; otherwise, it sends back an RST.

This new way of using cookies is more complex (and thus slower) than the original one, but has the advantage of allowing both client and server to negotiate and agree on TCP options that are relevant for performance. Thus, the performance penalty imposed by this 4-way handshake can be blurred by the performance gain in the subsequent data transfer. A similar 4-way handshake was adopted by the Stream

client socket state		segment		server socket state
1	CLOSED			LISTEN
2	SYN-SENT	→	(SEQ=*x*)(CTL=SYN)(tentative TCP options)	→
3	ESTABL.	←	(SEQ=cookie)(ACK=*x* + 1)(CTL=SYN,ACK)	←
4		→	(SEQ=*x* + 1)(ACK=cookie+1)(CTL=ACK)	→	ESTABL.

(I)

client socket state		segment		server socket state
1	CLOSED			LISTEN
2	SYN-SENT	→	(SEQ=*x*)(CTL=SYN)(tentative TCP options)	→
3	SYN-RCVD	←	(SEQ=cookie)(CTL=SYN)(final options)	←
4		→	(SEQ=*x*)(ACK=cookie+1)(CTL=SYN,ACK)(final options)	→	ESTABL.
5	ESTABL.	←	(SEQ=cookie)(ACK=*x* + 1)(CTL=SYN,ACK)(final options)	←

(II)

Figure 2.   Steps followed by TCP sockets using (I) the current SynCP implementation and (II) the basic approach of the new SynCP.

Control Transmission Protocol (SCTP [11]) to tackle the same security problem.

The TCP options are initially presented in the SYN of the client (step 2), and the final set of agreed options is returned in the server's SYN reply containing the cookie (step 3). The client's SYN-ACK (step 4) will simply reproduce the options presented by the server, as they already result from an agreement process; the same happens in step 5.

4.2.   Simplification of the basic approach

This basic approach can be further simplified: the final SYN-ACK sent by the server may be a simple ACK. Since client sockets are in a SYN-RCVD state, all they need to move to ESTABLISHED is an ACK. Consequently, we changed the protocol, replacing the SYN-ACK of step 5 by a simple ACK (see Figure 3).

client socket state		segment		server socket state
1	CLOSED			LISTEN
2	SYN-SENT	→	(SEQ=*x*)(CTL=SYN)(tentative TCP options)	→
3	SYN-RCVD	←	(SEQ=cookie)(CTL=SYN)(final options)	←
4		→	(SEQ=*x*)(ACK=cookie+1)(CTL=SYN,ACK)(final options)	→	ESTABL.
5	ESTABL.	←	(SEQ=cookie+1)(ACK=*x* + 1)(CTL=ACK)	←

Figure 3.   New SynCP with a final ACK instead of a SYN-ACK.

Early experiences showed that this simplification is not only possible but also critical. In fact, some TCP implementations follow a simplified state diagram where a socket in the SYN-RCVD state only changes to ESTABLISHED after receiving an ACK (Figure 6 of [3]), though they accept the SYN-ACK as a valid segment. This is a clear violation of TCP rules (c.f. [10], §4.2.2.10).

4.3. Initial assessment of problems

This new way of using cookies is a sort of Pandora box, since the exploitation of simultaneous connection initiations is rare, although valid and imposed by the seminal paper defining the TCP standard. Therefore, this protocol was tested and evaluated with several client operating systems to better assess its suitability. We tried to use both old and new systems, and also Unix/Linux, Windows and other proprietary systems (e.g. Cisco IOS).

Table 1. Client operating systems used to test the new SynCP and the result of a preliminary evaluation of their support for simultaneous connection initiations.

Operating System	OS or Kernel version	Supports the simultaneous connection initiation
Windows	CE 3.0 (PocketPC) 95, 98 SE, Millennium NT 4 Workstation/Advanced Server 2000 Professional/Server XP Professional	No
Cisco IOS	C4500-I-M V 11.1(7) C7200-DS-M V 12.0(7)	No
SunOS HP-UX Linux FreeBSD OpenBSD MacOS Digital UNIX OSF1 SGI IRIX	4.1.3, 5.6, 5.7, 5.8 A.09.05 2.2.x, 2.4.x 3.3 2.8 9.2.2 V4.0 5.2	Yes

Table 1 shows the exact systems that we experimented with and the preliminary results of using the protocols of Figures 2-II and 3. These tests showed two facts concerning the simultaneous connection initiation forced by the server:

1 Some operating systems apparently support it, but after a certain point they fail.

2 Some client operating systems support it, but react differently to the segments received.

Windows and Cisco IOS systems exemplify the first kind of systems. All the Windows systems tested fail the same way. After accepting the SYN-ACK reply, the client socket changes from SYN-SENT to SYN-RCVD, but thereafter it stays stuck in that state (repeatedly sending SYN-ACK segments to the server until giving up, sending then an RST; see Figure 4). We tried several possible replies to make it change state, including RST, but without any success.

The two Cisco IOS systems also fail but differently from the Windows systems. The client socket changes to SYN-RECV after receiving the

SYN, but replies with a simple ACK, instead of a SYN-ACK. Thus, from our point of view, these systems fail in handling the simultaneous connection initiation because we need to get a SYN-ACK segment from clients.

```
C.1711 > S.ssh:  S 3711264047:3711264047(0) win 64240 <mss 1460,sackOK> (DF)
S.ssh > C.1711:  SP 416925441:416925441(0) win 5840 <mss 1460,sackOK> (DF)
S.ssh > C.1711:  S 3878041854:3878041854(0) ack 3711264048 win 5840 <mss 1460> (DF)
C.1711 > S.ssh:  S 3711264047:3711264047(0) ack 416925442 win 64240 <mss 1460,sackOK> (DF)
S.ssh > C.1711:  . 1:1(0) ack 1 win 5840 (DF)
S.ssh > C.1711:  P 1:26(25) ack 1 win 5840 (DF)
C.1711 > S.ssh:  S 3711264047:3711264047(0) ack 416925442 win 64240 <mss 1460,sackOK> (DF)
S.ssh > C.1711:  P 1:26(25) ack 1 win 5840 (DF)
C.1711 > S.ssh:  S 3711264047:3711264047(0) ack 416925442 win 64240 <mss 1460,sackOK> (DF)
S.ssh > C.1711:  P 1:26(25) ack 1 win 5840 (DF)
C.1711 > S.ssh:  R 3711264048:3711264048(0) win 0
```

Figure 4.   Output produced by the tcpdump tool showing the segments exchanged using ssh in a Windows XP system to connect to the modified server using always the new SynCP. The PUSH flag in the second segment is explained in §4.6.3) and, for clarity, all NOPs of TCP options were removed.

In the second kind of systems we can distinguish two different reactions:

- Some only change to ESTABLISHED after receiving a pure ACK; getting a SYN-ACK only make them repeat their own SYN-ACK (e.g. SunOS 4.1.3). This behaviour, already referred to in §4.2, goes against TCP rules.

- Some acknowledge the SYN-ACK sent by the server, if using the protocol of Figure 2-II (e.g. SunOS 5.8). This is a legal behaviour.

These different behaviours show that our new SynCP is more sensitive to differences in TCP implementations of client operating systems than the current one. Though it may predict and accommodate, as much as reasonable, some known problems of client systems, there is always a possibility of failing with some of them.

4.4.   Overcoming problems raised by firewalls: mixed approach

Firewalls usually refuse TCP connections, initiated outside, to inside ports other than well-known service ports. This means that if the client of Figure 3 is behind a firewall, and the server is outside the defense perimeter of that firewall, the segment sent in step 3 will probably not reach the client. In that case, the client would continue to send SYN segments just like in step 2, until giving up. Therefore, the new SynCP will probably fail if the client socket is behind a firewall.

The solution that we devised for this problem is a best effort modification of the protocol presented in Figure 3. The modification consists of mixing both SynCPs, the current and the new, and thus the server replies to a SYN request with both a SYN and a SYN-ACK containing cookies. The format of the SYN-ACK is just like in the current SynCP, i.e. without any TCP options other than MSS. The server will try to deduce, from future segments sent by the client, which of the SynCPs it engaged to. Basically, if it receives a SYN-ACK, the client received the SYN and is using the new protocol; if it only receives an ACK, the client probably did not receive the SYN and is using the current protocol. If the segments with cookies sent by the server arrive in a different order to the client (the SYN-ACK first and the SYN next), the client will react to the SYN-ACK just like in the current SynCP (c.f. Figure 2-I). The delayed SYN will make the client reply with a harmless ACK (from Figure 10 of [3]).

Such mixing has a key issue, which is the compatibility between the SYN and SYN-ACK segments sent by the server. In practice this means that we have to decide if the cookies of these segments are the same or produced differently. Both solutions have advantages and drawbacks: equal cookies are natural to clients but may confuse the server; different cookies may be awkward to clients but allow the server to decide correctly. We chose the second approach, which is described below; for the sake of completeness, the problems raised by other approach are described in Appendix B.

4.4.1   Mixed SynCPs with two different cookies.

This approach simplifies the task of the server when dealing with probable segment losses because it knows exactly, from the acknowledged sequence numbers, which segments the client saw. The two cookies can be easily computed one from the other using a simple and fast invertible function, like a one's complement.

Its problem is that clients may react differently to the strange scenario of receiving two segments slightly incompatible between themselves. The main issue here is how should a client socket react when it receives a partially incorrect SYN-ACK, i.e. with an incorrect sequence number (server's ISN) and a correct acknowledged sequence number (client's ISN plus one).

According to [3], an RST should only be sent by a socket in any non-synchronized state (SYN-RCVD in this case) if "*the incoming segment acknowledges something not yet sent (the segment carries an unaccept-*

able ACK)". Since that is not the case, the normal reaction should be to either (i) ignore the segment, or (ii) send an ACK with the actual sequence numbers known by the client. In fact, our tests show that clients systems do react differently, but most of them send the expected ACK. Two systems, unfortunately, send RST segments: OpenBSD 2.8 and MacOS 9.2.2. This problem will be analysed further below.

So, the mixture of SynCPs using two different cookies – $cookie_1$ and $cookie_2$ (see Figure 5) – works this way:

   I If the client receives the server's SYN, then its socket, after changing to SYN-RCVD, waits only for an ACK to change to ESTABLISHED. The SYN-ACK can be received in the meanwhile, but because its sequence number ($cookie_2$) is different from the sequence number of the previously received SYN ($cookie_1$), it is invalid and an ACK is sent back to the server.

  II If the client misses the server's SYN because of a firewall, then it falls back to the current SynCP, as it only gets the server's SYN-ACK without any TCP options.

The server can easily check which of these alternative scenarios is the real one for each negotiation using cookies. If it gets a SYN-ACK with a $cookie_1$, then the client saw the SYN and engaged in the new SynCP. If it gets a simple ACK with a cookie, two scenarios are possible:

- **ACK acknowledges $cookie_1$:** the client saw the SYN and engaged in the new SynCP. The server simply drops the segment, as it should get a SYN-ACK with that cookie; the client will keep sending SYN-ACK segments until giving up or until getting a reply from the server.

- **ACK acknowledges $cookie_2$:** the client did not see the SYN and engaged in the current SynCP.

This mixed protocol using two cookies fails in two systems – OpenBSD 2.8 and MacOS 9.2.2 – because these, after accepting the SYN with $cookie_1$, do not reply with an ACK to the following SYN-ACK carrying $cookie_2$. Instead, they reply with an RST, which terminates the connection just established on the server side (see Figure 6).

However, such RST segments have some unusual properties that can help the server to detect and, possibly, overcome the problem. OpenBSD sends an RST but keeps the connection in the same SYN-RCVD state, which is an illogical reaction: if the RST is meaningful, it will eventually terminate the connection just created, so there is no point in keeping it. Fortunately, the RST is unusual and can easily be spotted and ignored

	client socket state		segment		server socket state
1	CLOSED				LISTEN
2	SYN-SENT	→	(SEQ=x)(CTL=SYN)(tentative TCP options)	→	
3	SYN-RCVD	←	(SEQ=cookie$_1$)(CTL=SYN)(final options)	←	
4		→	(SEQ=x)(ACK=cookie$_1$ + 1)(CTL=SYN,ACK)(final options)	⋯	
5		←	(SEQ=cookie$_2$)(ACK=x + 1)(CTL=SYN,ACK)	←	
6		⋯	(SEQ=x)(ACK=cookie$_1$ + 1)(CTL=SYN,ACK)(final options)	→	ESTABL.
7	ESTABL.	←	(SEQ=cookie$_1$ + 1)(ACK=x + 1)(CTL=ACK)	←	

(I)

	client socket state		segment		server socket state
1	CLOSED				LISTEN
2	SYN-SENT	→	(SEQ=x)(CTL=SYN)(tentative TCP options)	→	
3		✕	(SEQ=cookie$_1$)(CTL=SYN)(final options)	←	
4	ESTABL.	←	(SEQ=cookie$_2$)(ACK=x + 1)(CTL=SYN,ACK)	←	
5		→	(SEQ=x + 1)(ACK=cookie$_2$ + 1)(CTL=ACK)	→	ESTABL.

(II)

Figure 5. Steps followed by TCP sockets using an improved version of the new SynCP, capable of handling correctly clients behind a firewall. Scenario I shows the negotiation steps with a client not protected by a firewall; scenario II, on the contrary, shows the negotiation steps with a client protected by a firewall dropping pure out-in SYN segments. The values of **cookie**$_1$ and **cookie**$_2$ must be different and can be computed from each other using an invertible function.

I – OpenBSD 2.8

```
C.911 > S.ssh: S 1804448176:1804448176(0) win 16384 <mss 1460,sackOK,wscale 0,timestamp 12250862 0>
S.ssh > C.911: SP 3628984603:3628984603(0) win 5792 <mss 1460,sackOK,timestamp 1656630 12250862,wscale 0> (DF)
S.ssh > C.911: S 665982692:665982692(0) ack 1804448177 win 5840 <mss 1460> (DF)
C.911 > S.ssh: S 1804448176:1804448176(0) ack 3628984604 win 17376 <mss 1460,sackOK,timestamp 12250862 1656630>
S.ssh > C.911: . 1:1(0) ack 1 win 5840 <timestamp 1656637 12250862> (DF)
S.ssh > C.911: P 1:26(25) ack 1 win 5840 <timestamp 1656638 12250862> (DF)
C.911 > S.ssh: R 1804448177:1804448177(0) win 17376
C.911 > S.ssh: . 1:1(0) ack 26 win 17352 <timestamp 12250863 1656638>
S.ssh > C.911: R 3628984629:3628984629(0) win 0 (DF)
C.911 > S.ssh: P 1:23(22) ack 26 win 17376 <timestamp 12250863 1656638>
S.ssh > C.911: R 3628984629:3628984629(0) win 0 (DF)
```

II – MacOS 9.2.2

```
C.62884 > S.ssh: S 3091633285:3091633285(0) win 32768 <mss 1460,wscale 0> (DF)
S.ssh > C.62884: SP 2343418210:2343418210(0) win 5840 <mss 1460,wscale 0> (DF)
S.ssh > C.62884: S 1951549085:1951549085(0) ack 3091633286 win 5840 <mss 1460> (DF)
C.62884 > S.ssh: S 3091633286:3091633286(0) ack 2343418211 win 32768 <mss 1460,wscale 0> (DF)
C.62884 > S.ssh: R 3091633286:3091633292(6) win 0 (DF)
146.193.7.2.ssh > C.62884: . 1:1(0) ack 1 win 5840 (DF)
C.62884 > S.ssh: R 3091633286:3091633304(18) win 0 (DF)
```

Figure 6. Output produced by the `tcpdump` tool showing the segments exchanged using `ssh` in a OpenBSD 2.8 or a MacOS 9.2.2 systems to connect to the modified server always using the new SynCP with different cookies. The data in the RST segments sent by the MacOS consists of the following textual messages: "TH_SYN" and "No TCP/No listener". The PUSH flag in the second segment of each dump is explained in §4.6.3 and, for clarity sake, all NOPs of TCP options were removed.

by the server, enabling the mixed SynCP to be used with OpenBSD clients: from the dump of Figure 6-I we see that the RST includes a non-null window size, although this system usually sends RST segments like any other, i.e. with a null window size.

MacOS sends an RST and terminates immediately its ongoing connection. This RST is also unusual, because it includes data (see Figure 6-II), but that can only help the server to identify a client that does not support the mixed SynCP with different cookies.

4.5.   Final protocol: cache of problematic clients

Its now time to summarize all the problems faced by a mixed SynCP using different cookies in order to present a common solution for all of them. In short, the major problems are the following three:

- Some systems do not support the simultaneous connection initiation (e.g. Windows systems);

- Some systems do not react as required to the SYN sent by the server (e.g. Cisco IOS); and

- Some systems do not react well to a mixed protocol using different cookies (e.g. MacOS).

To handle all these cases we need to (i) maintain in the server a cache with the IP of problematic clients and to (ii) use only the current SynCP with hosts referred in that cache. Such cache should be updated whenever the server suspects a problem with the client. Furthermore, the cache should be managed in a conservative way, i.e. always assuming the worst case. This is advised because client systems may belong to private networks, using a gateway and masquerading to access Internet servers. In such cases, the server always sees the IP of gateways, but the protocol is sensitive to particular TCP implementations of client hosts behind them. Therefore, we should never remove hosts from the cache once they get there for some reason (except for getting a free entry).

The hints for inserting a client's IP in the cache are the following:

- The server receives a SYN-ACK, with a cookie, to a socket in the ESTABLISHED state. In this case we are probably dealing with a Windows client: we put its IP in the cache, but we don't abort the connection (first, because we may be wrong; and, second, because that is useless, as explained in §4.3); instead, the segment is processed normally by the TCP.

- The server receives an ACK, with a **cookie**$_1$, to a socket in the LISTEN state. In this case we may be dealing with a Cisco IOS client: we put its IP in the cache and we drop the packet.

- The server receives an RST, with the "TH_SYN" message, to a socket in the ESTABLISHED state. In this case we are probably dealing with a MacOS client: we put its IP in the cache and we let the RST be processed normally.

Note that in the first two cases the hint may be a false positive caused by: (i) a delayed reception of the server's ACK, in the first case, or (ii) a delayed client's SYN-ACK, in the second case. But, as previously explained, we should always assume the worst case; therefore we assume that such segments reveal a problematic client.

This cache is different from the one used by the Reset Cookies protocol to store security associations (c.f. §2). Both store the IP of real systems that tried to access the server, but our cache stores only the IP of problematic clients, while the other stores all the IPs. Thus, we are likely to get a better hit-rate with a cache of equal length. Furthermore, we only delay connections initiated by problematic hosts, while Reset Cookies delays the connections of all hosts not in the security association's cache.

The use of a cache of problematic clients is not a perfect solution, because the server reacts when it believes there could be a problem, instead of anticipating the problem. One possibility for an earlier detection of problematic clients could be to apply fingerprinting techniques, such as the ones used in active recognition tools (e.g. nmap [12]) or in passive IDS systems [13, 14], to the contents of SYN segments (either at TCP or IP level). This approach is not 100% accurate, may work better for some operating systems and may even be disturbed by fingerprint scrubbers [15]. Nevertheless, it may be explored in the future for some particular cases without interfering with the cache update policy previously described.

4.6. Security evaluation

4.6.1 Guessing SYN-ACK segments with valid cookies.

The mixed SynCP is as secure as the current one. Cookies are generated and validated the same way; they only appear in different TCP segments – in ACK segments in the current implementation and in ACK and SYN-ACK segments in the new one. The cookie of the SYN is computed from the one of the SYN-ACK using an simple and fast invertible function, like the one's complement. The fact of using two cookies instead of one does not reduce the resistance against guessing attacks, because at a given instant the set of cookies that is valid for a given type of segment, ACK or SYN-ACK, remains equal to that of the current SynCP.

4.6.2   Forged SYN segments with spoofed source addresses.

Another relevant concern with security is the impact of SYN segments sent by servers when replying to forged SYN segments sent by attackers. Unlike the current SynCP, that uses a normal reply, a SYN-ACK, the new SynCP uses a typical request segment (a SYN) as a reply to a client. This means that an attacker can lead a server under a SYN flooding attack to initiate connections with other servers. However, the algorithms to generate and validate cookies are enough to detect and avoid such problem.

Imagine the following scenario, illustrated in Figure 7: an attacker sends a forged SYN to a server A, which is using the new SynCP, and the forged segment says that the sender is an existing server B. The result of such attack is that A and B will exchange some segments and abort the connection, because the SYN-ACK from B has a cookie that was generated with x as ISN, and not with the ISN y provided by B in step 4. Furthermore, server B will also abort the connection by replying with an RST to any SYN-ACK segments sent by A to a socket in the LISTEN state (as in the current SynCP); such RST is produced by the normal operation of the TCP.

	server A socket state	segment			server B socket state
1	LISTEN				LISTEN
2		←	(SEQ= x)(CTL=SYN)	(apparently from B)	
3		→	(SEQ=$cookie(x)$)(CTL=SYN)	→	SYN-RCVD
4		←	(SEQ= y)(ACK=$cookie(x)$ + 1)(CTL=SYN,ACK)	←	
5		→	(RST)	→	CLOSE

Figure 7.   Segments exchanged resulting from a forged SYN referring an existing server B as the sender. The SYN-ACK segments with cookies that are also sent by A are not shown for the sake of simplicity, but they also abort the connection, because B replies with an RST to a SYN-ACK sent to a socket in LISTEN state (as in the current SynCP).

Note that the issue here is to avoid the creation of a useless TCP connection between A and B (between two sockets in LISTEN state) from a spoofed SYN segment sent by a attacker. Without using host authentication we cannot protect B from getting replies from A caused by spoofed segments. Neither can B prove that those segments were in fact sent by A.

4.6.3   Identification of SYN segments with cookies.   The
diagram of Figure 7 is not valid if both hosts A and B are servers acting

similarly, i.e. responding to SYN segments with other SYN segments carrying a cookie. In such a scenario, both hosts enter into an endless ping-pong of SYN segments, since they do not (intentionally) keep any record about past replies containing cookies.

This problem can be solved only if SYN segments containing cookies could be clearly distinguished from other SYN segments with ordinary ISN numbers. Two possible solutions for this problem are:

- to use one of the flags in the base TCP header not used in SYN segments (URG, PUSH, etc.); or

- to use a new TCP option.

The first solution is a sort of a hack that may work in most cases since TCP implementations are not sensitive to the state of such header bits in SYN segments. The second solution is more standard, all TCP implementations should be immune to it (see [10]) but it implies the reservation of a new option value.

Note that the clear identification of SYN segments is only needed for servers using the new SynCP, and not by any other hosts. Furthermore, such identification helps modified server hosts to further reduce the problem presented in Figure 7. In fact, as the host of server B can see that the segment from A is a SYN segment with a cookie, it may simply drop the segment and thus prevent all the following exchange of segments.

5. IMPLEMENTATION

The new SynCP, described in §4.4, Figure 5-I, was implemented on a Linux kernel (2.4.2-2). The implementation involved a minor modification of the TCP modules: three files (tcp_ipv4.c, tcp_input.c and syncookies.c) and about 300 new lines of code.

The implementation uses the following strategy for choosing SynCPs: if the client does not require any TCP options, or if the client belongs to our cache of problematic clients, the current SynCP is used; otherwise, we use the new mixed SynCP, described in Figure 5. To simplify the protocol tests, the kernel was also modified to behave as if under a SYN flooding attack.

The SYN segments used by the new SynCP are identified with the PUSH TCP header flag, as explained in §4.6.3. This flag was used in all the tests of the new protocol without any noticeable problems, but it should be replaced in the future by a proper, standard TCP option.

6.   CONCLUSIONS

In this document we presented a new strategy for using SYN cookies by a server under a SYN flooding attack. This new strategy overcomes a limitation of the current SynCP – it does not allow clients to negotiate any TCP options within SYN segments (it only allows clients to get the server's MSS). The solution that we propose relies on the fact that TCP allows a scenario called "*simultaneous connection initiation*", that we use to force client hosts to repeat their SYN requests. This way, the server can get together, in a single SYN-ACK, a cookie and all the TCP options initially requested by the client and already agreed to by the server.

This simple approach, fully compatible with standard TCP rules, faces two major problems. First, some systems do not deal correctly with the simultaneous connection initiation (e.g. Windows systems). Second, client-side firewalls may transparently interfere with the connection initiation started by the server, thus preventing the client from connecting to the server. To overcome these problems we did two complementary actions: (i) changed the protocol, in order to simultaneously use the current and the new SynCPs, creating a mixed SynCP, and (ii) added to the server TCP implementation a cache for storing the IP of problematic client hosts. This cache is updated whenever the server gets a hint, from the TCP segments received, that the client may not deal properly with the new mixed SynCP.

Concerning the security of the new protocol, we did not change the algorithms for generating and validating cookies, so they are as secure as they were before. We also showed that, due to the current algorithm to validate cookies, spoofed connection requests cannot drive a server to establish a connection with another victim server. Finally, we justified why SYN segments sent by the server must be properly identified to detect equal reactions of two hosts trying to connect with each other, both being under a SYN flooding attack. For simplicity we used the PUSH flag of the TCP header for such identification, without any noticeable problems, but a more correct implementation should use a proper, standard TCP option.

The new SynCP was implemented in a Linux kernel and tested with a large set of client operating systems. From the tests, we concluded that some systems do not tolerate it (Windows, Cisco IOS and MacOS), that some systems react strangely but in a way that can be detected and masqueraded by the server (OpenBSD), and that all the other systems behave as expected. The problems raised by the first kind of systems are solved with the cache of problematic systems.

In conclusion, we believe this new mixed SyncP, using both the current one and a new one faking a simultaneous connection initiation, is a valid and powerful improvement of the current SyncP. The resulting protocol supports the negotiation of any TCP options, is flexible enough to deal with firewalls and can be downgraded, on an as-needed basis, to the current one in order to attend to special problematic clients. In future implementations the late discovery of such clients may be partially anticipated by applying fingerprinting techniques to SYN segments.

Acknowledgments

I would like to thank Paulo Ferreira for reviewing draft versions of this paper and David Matos for reviewing the final version.

Appendix A: SYN cookies algorithms

Linux kernels use the folowing algorithms to generate and validate cookies:

$$H_1 = hash_{32-61}\left(S_{addr}|S_{port}|D_{addr}|D_{port}|K_1\right)$$
$$H_2 = hash_{32-61}\left(S_{addr}|S_{port}|D_{addr}|D_{port}|counter|K_2\right)$$

Generation:
$$cookie = H_1 + ISN_{client} + (counter \times 2^{24}) + (H_2 + data)\bmod 2^{24}$$

Validation:
$$counter_{cookie} = (cookie - H_1 - ISN_{client}) \div 2^{24}$$
$$\Delta counter = counter_{current} - counter_{cookie}$$
$$data = (cookie - H_1 - ISN_{client})\bmod 2^{24} - H_2 \bmod 2^{24}$$

$hash_n(x)$	n bit range, starting from $lsb0$, produced from x using the compression function of a digest algorithm (MD5 or SHA-1)
S_{addr}, S_{port}	source TCP/IP address
D_{addr}, D_{port}	destination TCP/IP address
K_1, K_2	secret keys
ISN_{client}	ISN provided by the client in the SYN segment
$counter$	minute counter
$data$	24-bit value

Cookies are generated and validated using two constant secret values, K_1 and K_2, which are long enough to completely fill the input buffer of the hash function used (64 bytes for both MD5 and SHA-1, so K_1 has 52 bytes and K_2 has 48 bytes). The data value is a server-defined value currently used for storing a 3-bit encoding of 8 predefined MSS values, presented in Table A.1.

The secrets K_1 and K_2 are produced using the kernel random number generator, the same used to generate the random part of ordinary IPv4 ISN values. These values are produced the first time SYN cookies are used after a system reboot and remain constant in kernel memory. The difficulty of guessing K_1 and K_2 from cookies is out of the scope of this document.

The kernel checks for suitable cookies only within a short time frame, starting when the last one was sent and ending a few seconds later. During that time gap cookies

Table A.1.   MSS predefined values encoded in the data value of SYN cookies.

SYN Cookie data	0	1	2	3	4	5	6	7
MSS value	64	256	512	536	1024	1440	1460	4312

may be checked, and are accepted only after some integrity control validations. There are two integrity controls of cookie contents, and if any of them fails the cookie is rejected. The first integrity control test checks if it is valid (acceptable) in terms of a temporal criteria: if $\Delta counter$ is lower than a given threshold (currently a hard-coded value of 4 minutes), it is acceptable. The second integrity control test checks if the data value is a valid one, i.e. a value between 0 and 7.

Appendix B: Problems raised by mixing SynCPs with two equal cookies

The main advantage of using equal cookies in both SYN and SYN-ACK segments used in a mixed SynCP is that clients always see segments that do not look strange. The instant chosen to send the server's SYN-ACK reply is irrelevant for the correction of the protocol from the client's point of view.

However, the premature sending of the server's SYN-ACK reply may be problematic for the server since it is not keeping state about ongoing connections. There are two particular scenarios that could lead to problems:

- The client socket receives the SYN and the SYN-ACK segments, sends replies whenever it decides to and moves to ESTABLISHED. If all the client's replies get lost, the client stays with a TCP connection that will be destroyed as soon as it sends some data or probes the server.

- The client socket receives the SYN and the SYN-ACK segments, sends replies whenever it decides to and moves to ESTABLISHED. If the server misses the SYN-ACK reply, but it sees one ACK reply, it will conclude that the client did not receive the SYN and, therefore, it uses the current SynCP. The result is that the server will establish a connection with the client, but will assume that the client will not use any TCP options, which is not true. This scenario can occur with client sockets that acknowledge SYN-ACK segments in the SYN-RCVD state, like SunOS 5.8.

The first scenario is annoying but not dramatic, being similar to a temporary server failure. The second scenario is more critical, since it can lead to future problems during the client-server interaction. However, it may be detected and avoided in some cases, namely when both client and server could agree on using TCP timestamps. In this case, the server could activate the time stamping in its SYN, but not in the SYN-ACK, and latter detect only from ACK segments if the client saw its SYN (if they carry a timestamp). This way ACK segments with both a cookie and a timestamp could not be used to create a connection.

Concerning the use of the TCP timestamp mechanism, the document describing it [9] says nothing about a segment not carrying a timestamp when the receiver is expecting it; it only says that timestamps may be sent only when the sender got one in the initial SYN of the connection. Therefore, we assume that it is legal to receive a SYN with a timestamp and a SYN-ACK without timestamp. Furthermore, such lack of timestamp in the SYN-ACK should not also affect the PAWS mechanism, also described in [9], because apparently it is only used for *"open connections"*.

References

[1] D. J. Bernstein. SYN cookies. http://cr.yp.to/syncookies.html.

[2] Syn cookies mailing list syncookies-archive@koobera.math.uic.edu. http://cr.yp.to/syncookies/archive.

[3] J. Postel. Transmission Control Protocol. RFC 793, September 1981. available via DDN Network Center.

[4] S. Bellovin. Defending Against Sequence Number Attacks. RFC 1948, May 1996. available via DDN Network Center.

[5] P. Ferguson and D. Senie. Network Ingress Filtering: Defeating Denial of Service Attacks which employ IP Source Address Spoofing. RFC 2267, January 1998. available via DDN Network Center.

[6] Livio Ricciulli, Patrick Lincoln, and Pankaj Kakkar. TCP SYN Flooding Defense. In *Comm. Net. and Dist. Systems Modeling and Simulation Conf. (CNDS' 99), 1999 Western MultiConf. (WMC' 99)*,, San Francisco, CAL, USA, January 1999.

[7] Eric Schenk. Another new thought on TCP SYN attacks, 1996. http://www.wcug.wwu.edu/lists/netdev/199609/msg00115.html.

[8] V. Jacobson and R. Braden. TCP Extensions for Long-Delay Paths. RFC 1072, October 1988. available via DDN Network Center.

[9] V. Jacobson, R. Braden, and D. Borman. TCP Extensions for High Performance. RFC 1323, May 1992. available via DDN Network Center.

[10] R. Braden. Requirements for Internet Hosts – Communication Layers. RFC 1122, October 1989. available via DDN Network Center.

[11] Q. Xie, K. Morneault, C. Sharp, H. Schwarzbauer, T. Taylor, I. Rytina, M. Kalla, L. Zhang, and V. Paxson. Stream Control Transmission Protocol. RFC 2960, October 2000. available via DDN Network Center.

[12] Fyodor. Remote OS detection via TCP/IP Stack FingerPrinting, October 1998. http://www.insecure.org/nmap/nmap-fingerprinting-article.html.

[13] Burak Dayıoğlu and Attila Özgit. Use of Passive Network Mapping to Enhance Signature Quality of Misuse Network Intrusion Detection Systems. In *16th Int. Symp. on Computer and Information Sciences*, November 2001.

[14] Honeynet Project. Know Your Enemy: Passive Fingerprinting. White Paper, January 2002. http://project.honeynet.org.

[15] Matthew Smart, G. Robert Malan, and Farnam Jahanian. Defeating TCP/IP Stack Fingerprinting. In *Proc. of the 9th USENIX Security Symp.*, 2000.

A MAC-LAYER SECURITY ARCHITECTURE FOR CABLE NETWORKS

Tadauchi Masaharu, Ishii Tatsuei, Itoh Susumu
Telecommunications   Hitachi,Ltd,.   Science University of Tokyo
Advancement Organization
of Japan

Abstract: Strengthened security is indispensable to the use of a cable network for e-commerce, the delivery of electronic public services, etc. This report proposes a new security system which better prevents tapping-related security violations. Enciphering in the system is on the Media Access Control (MAC) layer; including the MAC address among the enciphered items prevents the collection of information going to and coming from specific users by tapping. Simulation confirmed the operation of this system and its effectiveness.

Key words: cable network, MAC-layer, encryption protocol, symmetric cipher

1.   INTRODUCTION

Cable networks are installed for the distribution of video information. In the tree topology of these networks the head-end of the cable access television (CATV) system is the root and the customers are the leaves. A unique Media Access Control (MAC) address is assigned to each device that is attached to the network and information is distributed by tagging it with these addresses. This allows each customer terminal to receive the appropriate information from the head-end.

Encryption is indispensable to the provision of e-commerce facilities, electronic public services, etc., over such a network. However, catching the MAC address of a device allows the monitoring of information thus tagged. It will be possible to crack any encryption of this information given enough time. This allows tapping. We propose a new system that makes it impossible for a tapper to distinguish data according to the customer. The MAC address is simply included among the items that are enciphered—we call this the CNMS (Cable Network MAC-layer Security) system. We have confirmed the operation of the system by simulation; the simulation also implied that CNMS provides superior security against tapping.

2.   SYSTEM REQUIREMENTS

In considering a system to prevent the tapping of cable networks, we must start by considering the sources of this threat.

The tree topology of a cable network makes it possible to extract any information at any point in the network; it is easy to collect information going to and from the terminals of specific customers by using the MAC addresses and IP addresses that identify the terminals. These addresses should thus also be enciphered. The possibility of tapping remains, however, when a key is stolen, loaned, etc.

In table 1, required levels of security, i.e., of the use of cryptographic keys to prevent the cracking of tapped data, are classified into three levels according to the attack potential of aggressors which each must able to repel. A qualitative partition in terms of the attack potential is possible at level 2. This is discussed in section 5. There are fundamental differences between the offensive techniques employed by amateurs and professionals (pros): for example, an amateur is more likely to use monitored data to directly solve the encryption but a pro. tends to have a way of getting the key. A less talented amateur might also be able to receive a key. Obtaining a key is represented by level 1, the highest attack potential.

Table 1. Attack potential

Attack Level	Attackers	Notes	Definition
1	(Pro.) / Group	Pro: is given the means for the attack. Am.: takes advantage of situations.	The acquisition of a key. Defense is difficult.
2	Solo / Genius / Group	Capital and situation are the limitations here; all available techniques are used. The aims are tapping and imposture.	Propriety of information collection for decryption →differences in the required level of defense.
3	Solo (Am.)	From the technically naïve to engineers; attackers who are not particularly interested in tapping or imposture.	It monitors by setup which PC mistook.

Pro.: Professional, one whose occupation involves tapping and imposture.
Teams of thieves, agents from government organizations, detectives, etc.
Am.: Amateur, one whose occupation is independent of tapping and imposture.

Level 1: The attackers are able to obtain encryption keys;
Level 2: The attackers have tools for solving encryption keys; and
Level 3: One who is not especially interested in tapping and carries out no
   special action.

3. SYSTEM DESIGN

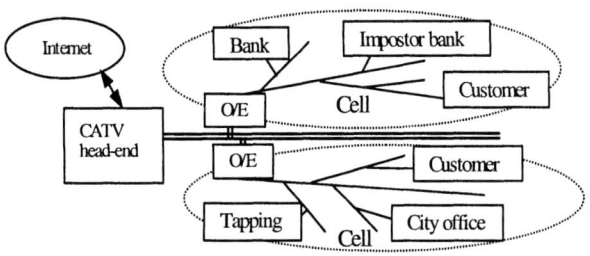

Figure 1. A cable network in the HFC configuration
(O/E: opto-electronic and electro-optic signal convertor)

The architecture of the CNMS system is described in this section. Fig. 1 shows an example of a cable network in the HFC (hybrid fiber/coax.) configuration. Several 'cells', each containing something in the region of 500 to 2,000 terminals, are connected with the CATV head-end by optical fiber. The tree topology within the cell is formed by coaxial cable.

A cell is set up for either symmetrical or non-symmetrical communication among its modems by the combination of the CMTS (cable modem termination system) at the CATV head-end and the CM (cable modem) operated by the customers. Upstream data and downstream data are always assigned to separate frequency bands. In a symmetrical modem system, upstream data are frequency-converted into downstream data before they are input into the CMTS at the CATV head-end. This allows peer-to-peer communications among terminals and sets of terminals within the cell. The target and originating terminals are represented in data flowing downstream and upstream, respectively, and this information is easy to intercept.

In the case of non-symmetry, communications are one-to-N from the CMTS to sets of N CMs and N-to-one from the sets of N CMs to the CMTS. An asymmetrical modem is described in DOCSIS 1.0 or 1.1 [1]. The downstream information on the network is still distributed to all of the terminals in a cell. Realizing encryption that encompasses the MAC address in a symmetrical modem-based system complicates the management of keys in the CM. Implementation is thus difficult. The CNMS proposed here is for systems based on asymmetrical modems.

3.1   Overview of the CNMS

The CNMS (Fig. 2) adds a security function to asymmetrical type modems such that encryption is realized in the MAC layer. In the PDU (Protocol Data Unit) format, the destination source MAC addresses are included among the items enciphered by the CNMS. This prevents the identification by tappers of specific subscribers as targets.

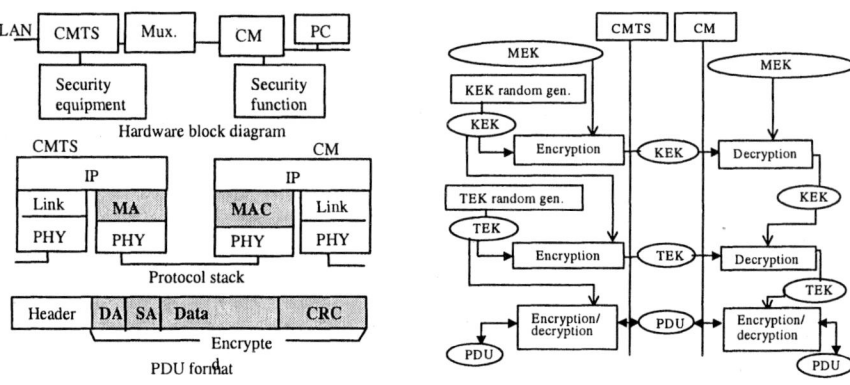

Figure 2. Concept of the CNMS   *Figure 3.* Arrangement of keys for the CNMS

Fig. 3 is a schematic diagram of the arrangement of keys at the CATV-head and customer ends when using symmetric cipher keys in carrying out an encrypted communication. The master encryption key MEK is placed as the symmetric cipher at both ends of the communication and is used to encipher and decipher the randomly generated key- and traffic-encryption keys, KEK and TEK. The TEK is used to encipher the transmission. All data within the cable network are enciphered in this way. Moreover, each key xEK has a lifetime L which follows L(MEK)>L(KEK)>L(TEK).

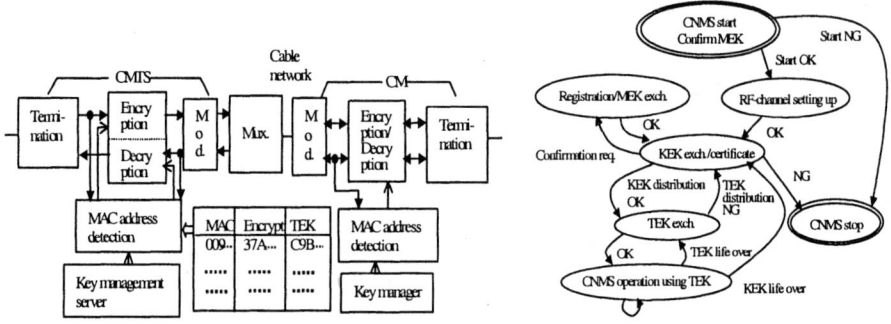

Figure 4. Block diagram of the CNMS   *Figure 5.* State transition diagram
                                                        for the CNMS of a CM

3.2 Configuration and Operation of the CNMS

A block diagram of the proposed CNMS system is given as Fig. 4. A single key is in use in the cable modem (CM) of each customer; this allows satisfactory performance in decryption even with data sent at 30 Mbps. The CATV head-end, however, holds the keys for many customers, and having it individually and sequentially decrypt all of the data is not practicable. The MAC addresses as encrypted by the respective CM keys at a given time are thus memorized in a table which is part of the MAC address detection equipment of the CMTS at the CATV head-end. When a MAC address as encrypted by a CM is received at the head-end, the CMTS refers to this table to immediately identify the key. Fig. 4 shows the case where the encrypted MAC address is immediately detectable by both the CMTS and CM.

The most important key information is held by the CNMS. Key information is accumulated by a key-management server at the CATV head-end and by a key manager at the CM. Two cases are the saving of the master encryption key for a CM, and the case where a key is placed in an IC card, while another is set in a cable modem (CM). IC cards for BS broadcasting are to be mounted in the set-top boxes (STBs) of CATV systems, so the sharing of IC cards will be desirable for unified systems.

Fig. 5 is the state transition diagram for the CNMS of a CM. Here, the CMTS at the CATV head-end is already operating. This state transition diagram becomes effective when power is suppled to the CM.

The first transactions after the CNMS has started up is a check for the presence of an MEK in the CM and setting up of the RF-channel asssumptions [1] for the CM.

The CM's initial CNMS state is then processed. The coincidence between the MAC address and MEK on the CM and the values registered at the CATV head-end is checked; if they match, the process of KEK exchange/authentication proceeds.

At this time, the conditions for suspending CNMS operation include authentication failure and CNMS-start failure. When the setting-up of the RF-channel for a CM fails, the operation of that CM is stopped. When the TEK that has been encrypted by KEK after the successful distribution of the KEK is itself successfully distributed, encrypted communication using the TEK starts up. When the life-time of the KEK expires or TEK distribution is not possible, the system's state returns to the exchange of the KEK. CNMS operations stop if CNMS operation using the TEK is impossible even after several attempts to distribute the KEK and TEK. After the CNMS has been stopped, operations continue in a mode in which encryption is not employed.

3.2.1   CNMS Initialization

Let the period from start-up of the CNMS to operation of the CNMS with the TEK be the period of CNMS initialization. The CNMS and its initialization must be appropriate for a system based on asymmetrical modems. According to versions 1.0 and 1.1 of the DOCSIS specification, the initialization of the CM is as follows:
1) scannning and synchronization in the downstream direction;
2) obtaining the upstream parameters (UCD: upstream channel descriptor);
3) range-finding and automatic adjustments;
4) establishing the IP connections (DHCP: dynamic host-configuration protocol);
5) establishing the Time of Day;
6) transfer of operational parameters; and
7) registration for a CM.

There are three points at which the CNMS may be initialized; between steps 2) and 3) (case one), between steps 3) and 4) (case two), and after the registration step 7) (case three). The three cases and their good and bad points are summarized in table 2.

Table 2 CNMS initialization

Case	Techniques	Merits	Demerits
1	The MEK is used to encrypt the ranging information.	All of the information in the cable network is enciphered.	The prior registration of the CM's MAC address in the CMTS is indispensable (auto-registration is not possible)
2	The CNMS is set-up after ranging.	IP addresses are enciphered.	The timing of communications in equipment authentication is clearly known.
3	The CNMS is set-up after registration of the CM.	Initialization of the asymmetrical modems is easy.	The MAC and IP addresses that are in use become clear. The timing of communications in equipment authentication is clearly known.

Although per-customer tapping is clearly most difficult when the CNMS start-up is earliest, this approach places limits on the system's functions; e.g., in the first case, for the MEK to be used in encrypting the ranging demand from the CM, the MAC address and MEK of the CM have to already be in the CATV head-end. When users install new modems, they have to connect the modems to the CATV head-end at the MAC address level. Making a connection at the MAC address level is not easy for an amateur. On the other hand, in the second and third cases, customer authentication on the basis of the MEK is possible from the times of ranging and registration, respectively, and automatic registration of the MAC address is possible. The second and third cases differ in whether or not someone who is tapping the network is able to read the IP addresses of the CMs. Moreover, the vendor ID of the modem and the MAC address of the CPE (customer-premise equipment), etc., are all clear when the CM is registered. After CNMS operations have commenced, the MAC address and IP address are encrypted, so the IP address is not distinguishable. The level of security is thus not strongly affected by the shift from the first to the second or third case. The problem of the timing with which the MEK is applied is thus solvable, to some extent, by having the MEK as the key for a strong form of encryption such as 3DES.

3.2.2 The Distribution of the TEK

All of the keys used by the CNMS, i.e., the MEK, KEK and TEK, are symmetric cipher keys. Of these keys, the MEK is mounted in an IC card or the CM and is delivered from the CATV head-end to the customer. Avoiding the use of a public key as the MEK obviates the authentication of a public key; the possession of the symmetric key itself is used for authentication. This simplifies the system. A further advantage of this approach is that delivery of the key should be comparatively easy in a fixed network such as a cable network.

New TEKs are frequently distributed but the periods between the distribution of KEKs are comparatively long. The TEK is changed at the end of a short and fixed period or on every transmission of data from the CM. This ensures that the encrypted MAC address is represented by different data on each transmission or in each of the short periods.

When the TEK is changed every time a fixed period elapses, the MAC address may be transmitted more than once during one period, i.e., as the same data. Let the fixed period be *tex* seconds. If n TEKs with attached index numbers are transmitted to each CM from the CMTS every time the keys are changed, the CM is able to randomly select one of the TEKs, attach its index number to a header, and transmit the result as encrypted data to the CMTS. In this case, a new set of TEKs is transmitted from the CMTS every $(n \times tex)$ seconds, and in response to this, the CM returns encrypted Ack data to CMTS. Although an asymmetrical cable modem is able to handle a downstream rate of 30 Mbps and thus has a large capacity, since the maximum upstream rate is about 10 Mbps, we need to look at whether or not the upstream flow of traffic is affected by TEK distribution. If N CM units are connected in one cell of a cable network and N is 1,000, the period between changes of the TEK *tex* is four seconds, the number of potential TEKs sent n is 4, and the Ack data *Dak* consists of 64 bytes, the upstream data rate for TEK exchange is

(*Dak* x 8) x*N* / (*n* x*tex*)=32 Kbps.

Although the upstream capacity will generally be much greater than 32 Kbps, the TEK-exchange period *tex* may be extended when sufficient capacity is not available.

3.3 System Comparison

Table 3 System comparison

Item	CNMS	BP
MAC address	Encrypted with the data.	Not encrypted.
IP address	Encrypted with the data.	Encrypted, but not during IP setup.
Authentication	Based on both the MAC address and MEK.	Based on the MAC address of the CM and the user's RSA secret key.
Collection of information on a specific customer	Extremely difficult.	Possible (however, the data is encrypted).

The specification of Baseline Privacy (BP) [2], which is managed by U.S. Cable Labs, includes the use of RSA public keys in the changing of the symmetric cipher key MEK for a CNMS, and the difference is making to solve with the application of a hash function to get the KEK for Baseline Privacy. However, in Baseline Privacy, encryption is not applied to the MAC addresses (DA and SA), and Baseline Privacy does not apply encryption when the IP addresses are set up. Someone who sees a MAC address at this stage is able to detect the terminal to which the information is being sent and its IP address. As is shown in table 3, a comparison of the results of applying the BP and the CNMS, the collection of information on a specific user is more difficult with the CNMS. Since several alterations of the TEK per day are sufficient for Baseline Privacy, the effect of these alterations on the upstream flow of traffic is negligible .

4. THE EXPERIMENT BY SIMULATION AND ITS RESULTS

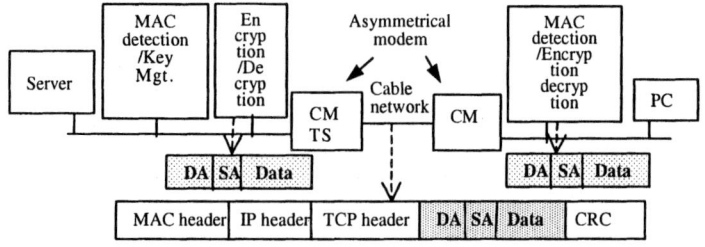

Figure 6. Block diagram of the experimental system

Although further experiments with the CNMS system will be necessary, including full testing on an actual CMTS at the CATV head-end and the CM units of customers, we have carried out an experiment by simulation at the application level. The simulation is of a limited environment.

4.1   Block Diagram of the Experimental System

The block diagram of the experimental system (Fig. 6) shows the encryption/decryption equipment and MAC-address-detection/key-management server that are added to the asymmetrical cable modem termination system at the CATV head-end and the MAC address-detection equipment, which includes a key manager, that is added to a customer's CM. The PC and the server make data, to which MAC addresses (DA and SA) are added, and in each case the whole is then encrypted. This encrypted data is then bi-directionally transmitted after a session has been established between the PC and the server. If both sets of encryption-decryption equipment correctly decrypt the encrypted data that has been sent, the CNMS is realizable.

4.2   Experimental Results

Logs of data that are the results of simulation with eight PC/CM sets, with TEK being altered every second, a downstream rate of 10 Mbps and an upstream rate of 2 Mbps, are given in Figs. 7 and 8. Each item of encryption/decryption equipment and the MAC-detection/key-management server ran on a 700-MHz CPU. The upper part of Fig. 7 shows the results of decryption of the MAC address in the log of data (lower lines) and the results of encryption (upper lines).

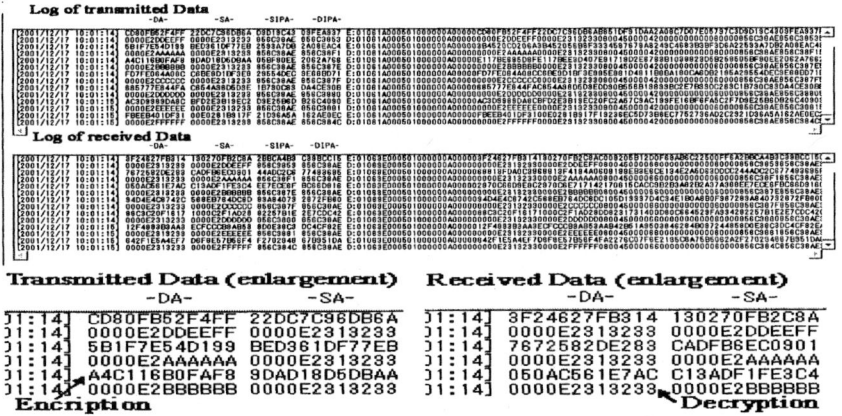

Figure 7. Logs of Data at the CATV head-end

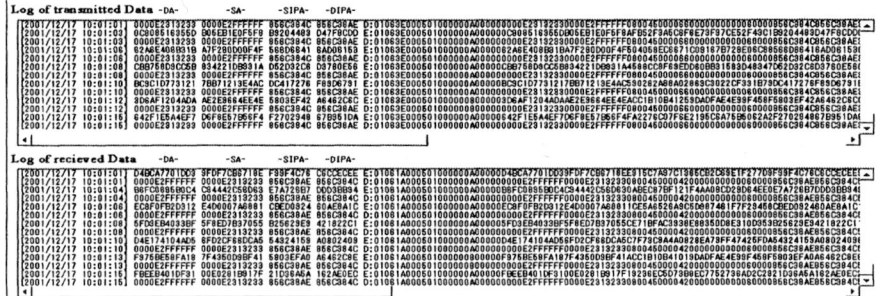

Figure 8. Logs of Data in a customer's PC

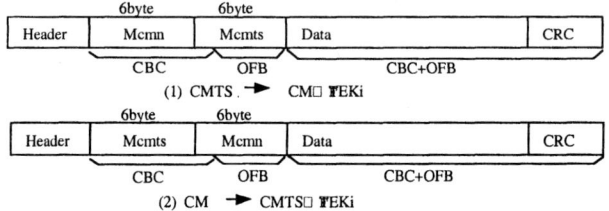

Figure 9. Method for encryption of MAC addresses

In Fig. 8, only the MAC address and IP address that correspond to the customer's CM have been obtained. The results of encryption for data transmitted to and received from a single CM at the same time are not the same, despite both transmissions occurring during a single TEK period (Fig. 7, lower part). Fig. 9 shows us why; the sequences with the MAC address of the CMTS, Mcmts, and the sequences with the MAC addresses Mcmn of the CM units are in opposite order in data transmitted and received by the CM, and the 8-byte block cipher MULTI2 includes 12 bytes of encrypted data in its CBC(cipher block chaining) and OFB(output feedback) modes. Sets of these 6-byte encryption results are shown in each of the enlarged displays of Fig. 7, and DA and SA head the corresponding columns. The same value is used for the initial vector in the cipher in CBC and OFB modes by all CMs and CMTS.

5.   SYSTEM EVALUATION

The experiment result of Section 4 shows that the source and target of data are not identifiable in a system where the CNMS is applied. This provides security against tapping. To obtain data for a particular target, a person who is tapping the network needs to be able to memorize all of the data in the 30-Mbps stream, and then to solve the encryption for each of the segments. It is difficult to continue memorizing all of the data and then encrypt it quickly enough for the result to be useful. Baseline Privacy (BP), on the other hand, allows the identification of data for individual target customers; a tapper is thus able to collect data for a given customer and then solve the corresponding encryption. Moreover, when the RAS secret key is also known, the tapper is able to decrypt all of the data. This also applies when IPSec [3], SSL [4], etc. is used with the cable network, i.e., as long as the MAC address is not included among the enciphered items. When the star topology is used to connect the central office with the individual customers, as is the case for telephone lines, it is impossible for a customer to obtain the data for others on the network without tapping the corresponding lines. When a wiretap is connected to a circuit, even if the addresses are encrypted, the person running the tap will be able to collect a specific customer's information. The installation of wiretaps by professional snoops and thieves always remains a possibility.

As an aside, no practically applicable security technique is able to handle attacks at level 1 of table 1. Although the quantum cipher makes it possible to detect interception, it is still difficult to prevent tapping. Even though attacks at potential level 3 require no special measures, a symmetrical modem allows a user to view common files that have been carelessly set up by other customers. The above points allow us to summarize robustness of security against tapping in table 4. We assume equal security for all ciphers.

Table 4 Robustness of security against tapping

Attack potential	Technique	Points
Level 1		
High	CNMS	All addresses are enciphered. Collecting a specific customer's information is very difficult.
↑ Level2	Telephone-line +IPSec, SSL, etc.	Wiretapping is an effective way to acquire data.
↓ Low	BP, CATV+IPSec or SSL	The MAC address offers a good way to collect the information of specific customers.
Level 3	Asymmetrical modem	Tapping of downstream data is possible.
	Symmetrical modem	Tapping of downstream and upstream data is possible.

6. CONCLUSIONS

We have proposed the CNMS, where enciphering is in the MAC layer and thus applied to the MAC address, as a way to improve the security of cable networks. The CNMS destroys the value of tapping, which is the weak point of cable networks, and so it is more secure than other security systems. We confirmed the promise of the CNMS in an experiment by simulation of an asymmetrical-modems-based cable network. We confirmed that the system provides a practicable solution. A standard is indispensable to widespread application of the CNMS. We thus need to look into the optimization of the cipher protocols, etc.

REFERENCES

[1] DOCSIS (Data-over-Cable System Interface Specifications): "Radio Frequency Interface Specification", SP-RFI-I06-010829, August 2001

[2] DOCSIS (Data-Over-Cable System Interface Specifications): "Baseline Privacy Interface Specification", August 2001

[3] RFC2401 "Security Architecture for the Internet Protocol", November 1998

[4] RFC2246 "The TSL Protocol Version 1.0", January 1999

TOWARDS AUTHENTICATION USING MOBILE DEVICES
An Investigation of the Prerequisites

E. Weippl, W. Essmayr, F. Gruber, W. Stockner, T. Trenker
Software Competence Center Hagenberg

Abstract: In this paper we show how mobile devices can be used as authentication tokens. We highlight the prerequisites such as mobile device security and mobile communication security. We elaborate on already existing solutions and on what issues in the context of security remain to be addressed. Beside the comprehensive overview, our main contribution is to explain how the different characteristics of wireless communication can be abstracted. Based on this abstraction an implementation of mobile authentication is transparent both to the application programmer and to the end users.

Key words: security, mobility, authentication

1.   INTRODUCTION

Business analysts predict great strategic and technical prospects to upcoming mobile business applications using short distance as well as wide area wireless communication facilities. Among others, the most striking benefit is that information may "really" be accessed at any time, anywhere, and with any device, available.

A very promising service that can be delivered to mobile users is that of authentication - referred to as mobile authentication. Since mobile devices are portable, personal, useful, and valuable, users tend to carry them everywhere they go. A number of applications are feasible based on mobile authentication such as login, logout, locking, and unlocking a computer or signing in on attendance lists etc. However, the acceptance of such mobile services will strongly depend on the degree of trust the user can have in mobile and wireless facilities thus making their security decisive for mobile authentication. In this paper we investigate the prerequisites for mobile authentication.

2.   PROPERTIES AND APPLICABILITY OF MOBILE DEVICES FOR SECURITY SERVICES

This section discusses the main characteristics that are relevant when using these mobile devices for security services.

2.1   Mobility

Since devices are small and portable users tend to carry them along most of the time, at work and after office hours. This makes the devices an ideal platform to implement additional security services such as authentication. Unlike with key cards users are not required to carry an additional item that serves one purpose only. Beside all the functionality available, the devices are programmable. Therefore various degrees of authenticity can be implemented offering also protection of the user's privacy.

Privacy concerns are especially important as mobile devices connect to other services using wireless communication. Wide area wireless networks allow users to use Web services independent of their current location. Wireless LANs (WLANs) or Bluetooth allow to connect to local services, which makes it well suited for local authentication services such as unlocking doors or signing in on attendance lists. One drawback of technologies is that it is difficult to establish the intent of the user merely by his/her presence.

2.2   Value

Mobile devices are rather expensive. They are valuable not only because of the price of the device but also because they store private data. Nonetheless most devices lack effective authentication mechanisms; anyone who possesses the device can use it. Cell phones use PINs but they are often only required when switching the phone on. In most cases the devices will constantly remain on and thus users are never authenticated to the mobile device again.

Since the devices are valuable people always carry them along; thus the devices are extremely well suited for applications that identify people to services.

2.3 Personal Property

As previously mentioned cell phones and PDAs are personal devices that people do not share. Unlike [Eckert, 2000] we therefore do not believe that multi-user support on PDAs is essential. In future such devices could supersede traditional keys or ATM cards. To guarantee that mobile devices really remain personal, strong and convenient authentication towards the device is required. This paper does not address the issue of how a user authenticates to the device but how the device can be used to authenticate the user to other services.

2.4 Security and Usability

Based on the aforementioned arguments, mobile authentication promises to increase both usability and security. Users of mobile devices should not have to be concerned about and should be safe from threats while using their device for secure actions. Furthermore, an increased subjective perception of security will foster a broader spectrum of use cases and thus also increase usability.

Moreover, usability will inevitably be increased, as users will not have to carry additional devices; instead, their mobile devices will also perform authentication services. Generally it does not depend on what kind of device is used. Programmable devices will have a greater chance of fulfilling the need of providing different levels of security.

Security obviously involves both the device itself and the (wireless) connection of the device to its environment. In the following sections we first elaborate on the security of the mobile device and second on wireless communication security.

3. MOBILE DEVICE SECURITY

Figure 1 illustrates a typical mobile environment in which mobile devices are used for conducting mobile business.

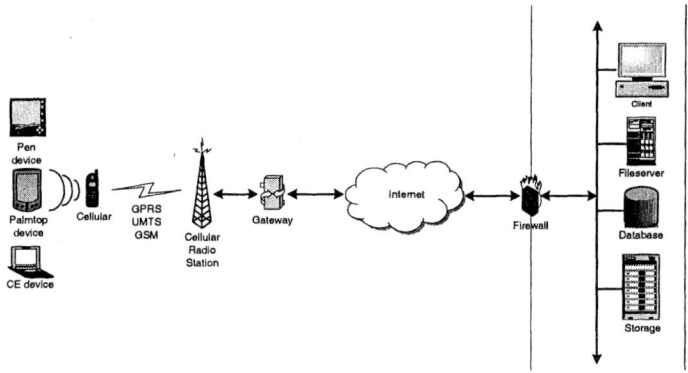

Figure 1: Typical Mobile Business Environment

The mobile business environment includes mobile devices, communication infrastructure, multiple wireless communication protocols, Internet related communication protocols, and different (corporate) networks with all its components such as firewalls, backend systems, or client applications. Within the scope of this paper we do not concentrate on corporate network or Internet related security issues, since these topics are well covered in literature. Instead, we provide an overview of the prerequisites for our vision of mobile authentication as described in section 5, namely, mobile device security (covered in the remainder of this section) and wireless communication security (covered in section 4).

3.1   Security Risks of Mobile Devices

The security risks particular to mobile devices result from the typical properties inherent to such devices, namely, they are personal, portable, have limited resources and are used for roaming through networks with different levels of trust.

Mobile devices are felt to be highly personal equipment inducing their users to have increased trust and manage confidential private and corporate information or even conduct security relevant actions like payment, for instance.

The portability makes mobile devices subject to loss or theft. If a mobile device has been stolen or lost, unauthorized individuals may gain direct access to the device's resources. Furthermore, the identity of the device's owner may be claimed when roaming through the owner's networks and the resources there might be exploited. In the worst case, devices could be faked and accidentally returned to the user who from now on unconsciously delivers valuable information to the potential attacker.

The limited resources of mobile devices increase the probability of denial-of-service risks. For instance, if a cell phone can be used for authentication, a power failure suffices to prevent legitimate access. Unfortunately, the current practice when addressing resource limitations is to ignore well-known security concepts. For instance, to empower WML scripts, implementations lack the established sandbox model and downloaded scripts can access local resources without restriction [Gosh, 2001].

Moreover, roaming through different cells and networks with varying degrees of trust and different security policies renders current techniques such as SSL and even protocols specifically designed for mobile use (e.g. WTLS) vulnerable to new forms of attack [Gosh, 2001]. The mobile device has to be compliant to multiple security policies and frequently adapt to different trust levels found within the visited networks. Furthermore, the user might be subject of undesired tracing and profiling due to the location awareness of particular devices, which can lead to substantial privacy problems.

3.2   Security Services

To address the security risks identified within the previous sub-section, a number of security services are used. Within an industry project we searched for security enhancements available for PalmOS and WinCE devices according to our partners' requirements.

3.2.1   Physical Protection

Steel cables, holsters, etc. to fix the mobile device on a person's body or e.g. a computer chassis provide protection against theft, loss and damage. They allow attaching the mobile device to a neck strap or key chain. Several products for physical protection of Palm handheld devices are available such as Bond, e-Holster, or PDA Saver.

3.2.2   Authentication

Authentication on the mobile device establishes the identity of the user to the particular mobile device. Most of the available mobile devices only provide basic authentication using personal identification numbers (PINs) or passwords. Frequently, users may also turn off authentication for their convenience. Thus, more convenient and more secure authentication mechanisms will be needed when using the mobile device for mobile authentication as suggested in this paper. These mechanisms should at least be based on something the user *has* (e.g. a smart card that must be inserted into the mobile device for critical actions) or something the user *is* (exploiting a biometric signature such as a fingerprint for authentication).

At least some products are yet available that provide prominent PDAs with enhanced authentication features. For instance, PDASecure (PalmOS) and Sing-On (PalmOS, WinCE), that enable password encryption or PINPrint (PalmOS, WinCE) that provide fingerprint authentication.

3.2.3   Access Control

Based on a legitimate identity proved by authentication, the mobile device should further restrict access to its resources. Although a PDA will mostly be used only by its owner, there might also be the need to share such devices in some cases. Most of the mobile devices do not provide any access control at all. For PalmOS, some products (e.g. Enforcer, Restrictor) are available that provide profiles to limit access to specific data.

3.2.4   On-Device Encryption

Authentication and access controls may not suffice to protect highly sensitive corporate or private data stored on the mobile device. It is often necessary to provide a redundant level of protection by encrypting all or parts of the devices data storage. Several products can be installed in order to get on-device encryption, such as JawzDataGator or MemoSafe for PalmOS, and CryptoGrapher for confidential information kept on flash cards, PocketLock for encrypting documents, or seNTry 2020 for protecting volumes, files, folders and programs all of which can be used on WinCE devices.

3.2.5   Anti-Virus issues

Anti-virus software should be used to protect both the handheld and the (corporate) network against malicious software. Products that support anti-virus protection for mobile devices are, for instance, InoculateIT and Palm Scanner for PalmOS respectively VirusScan, and Anti-Virus for WinCE.

3.2.6   Application Security

It is not enough to secure the mobile device itself and the wireless communication protocols as reported in section 4. Also the applications running on the mobile device have to be secured. Applications should be designed enabling authentication, authorization, access-control, and encryption mechanisms that can operate across platforms and technology domains. At least some products supporting basic cryptographic algorithms and applied cryptographic mechanisms like SSL, for instance, are available (e.g. Security Toolkit for PalmOS or Security Builder for PalmOS and WinCE).

4.   WIRELESS COMMUNICATION SECURITY

Security does not end at the device, thus a general security concept including wireless media must be developed. In the last section we described the security aspects of the mobile device itself. This section describes in short terms special security threats when dealing with wireless communication technologies. In a subsection we will summarize the security aspects, which the existing wireless technologies have already built-in. Moreover we will show potential weaknesses of each communication technology.

4.1   Wireless Security Threats

The following security threats are not particularly special for mobile computing, but with wireless communication technologies certain new aspects arise, which we will describe in the following enumeration (compare [SANS, 2001] and [Eckert, 2000]).

– Denial of Service (DoS) occurs when an adversary causes a system or a network to become unavailable to legitimate users or causes services to be interrupted or delayed. A wireless DoS attack could be the scenario, where an external signal jams the wireless channel. Up to now there is little that can be done to keep a serious adversary from mounting a DoS attack. A possible solution is to keep external persons away from the signal coverage, but this is rarely realizable.

– Interception has more than one meaning. An external user can masquerade himself as a legitimate user and therefore receive internal or confidential data. Also the data stream itself can be intercepted and decrypted for the purpose of disclosing private information. Therefore some form of strong encryption as well as authentication is necessary to protect the signals coverage area.

– Manipulation means that data can be modified on a system or during transmission. An example would be the insertion of a Trojan horse or a

virus on a computer device. Protection of access to the network and its attached systems is one means of avoiding manipulation.
- Masquerading refers to claiming an authorized user while actually being a malicious external source. Strong authentication is required to avoid masquerade attacks.
- Repudiation is when a user denies having performed an action on the network. Strong authentication, integrity assurance methods and digital signatures can minimize this security threat.

4.2   Wireless Communication Security

In this section we describe how the various communication technologies try to overcome the potential security issues mentioned above.

4.2.1   Wireless LAN (IEEE 802.11)

Wireless LAN (WLAN) specifies two security services; the authentication and the privacy service. Mostly these services are handled by the Wired Equivalency Privacy (WEP). WEP is based on the RC4 encryption algorithm developed by Ron Rivest at MIT for RSA data security [RSA, 2002]. RC4 is a strong encryption algorithm used in many commercial products. The key management, needed for the en/decryption is not standardized in WLAN but two key-lengths have come up: 40bit keys for export controlled applications and 128bit keys for strong encryption. Some papers on the uncertainty of the WEP standard have been published [Borisov, 2001] but [Kelly, 2001] from the 802.11 standardization committee responded in the following way: WEP was not intended to give more protection than a physically protected (i.e. wired) LAN. So WEP is not a complete security solution and additional security mechanisms like end-to-end encryption, Virtual Private Networks (VPNs) etc. need to be provided.

4.2.2   Bluetooth

In the Bluetooth Generic Access Profile (GAP, see Specification [Bluetooth, 2002]) the bed-rock on which all other profiles are based, 3 security modes are defined:
- 1: non-secure
- 2: service level enforced security
- 3: link level enforced security

In security mode 1 a device will not initiate any security - this is the non-secure mode. In security mode 2 the Bluetooth device initiates security procedures *after* the channel is established (at the higher layers), while in security mode 3 the Bluetooth device initiates security procedures *before* the channel is established (at the lower layers). At the same time two possibilities exist for the device's access to services: "trusted device" and "untrusted device". The trusted devices have unrestricted access to all services. The untrusted device does not have fixed relationships and its access to services is limited. For services, 3 security levels are defined: services that require authorization and authentication, services that require authentication only and services that are open to all devices. These levels of access are obviously based on the results of the security mechanisms themselves so they are not really of interest to us. Thus will concentrate on the two areas where the security mechanisms are implemented: the service level and the link level. Details on how security is handled in these levels can be found in [Palowireless, 2001].

Though Bluetooth has a large concern on security, Bluetooth is not free from vulnerabilities, what is shown in following papers and web sites [Vainio, 2000; Sutherland, 2000] and so additional security and encryption must be provided for an acceptably secure data transmission.

4.2.3   Infrared Media

The infrared communication technology does not provide any security mechanism incorporated into the transmission protocol. The standardization committee, the Infrared Data Association (IrDA, [IRDA, 2002]) justifies this with the very limited spatial range of infrared media and with the required sight-to-sight connection of the involved devices.

4.2.4   Wireless Wide Area Networks

The security of digital wireless WANs (GSM, GPRS, HSCSD, etc.) is described in this section in an abstract manner, taken from [SCCH, 2001].

Following [Walke, 2000] and [Hansmann et al., 2001], the user of a network must on the one hand be identified for billing purposes; on the other hand the transmitted data must be protected for privacy reasons. For digital wireless WANs cryptographic methods are used to implement the security features.

The first level of security concerns the mobile device. When a device is manufactured, a unique ID is assigned to each mobile device. When a user wants to access a network, first the ID of the used device is sent to the network. A device can be classified into three categories:

– White-listed: Access is allowed,
– Gray-listed: Access is allowed, but mobile device is under observation, and

– Black-listed: Access is not allowed (e.g., mobile device is reported stolen)

In a next step the user that wants to access a network is identified. Each subscriber to a network is issued a unique security key and a security algorithm. Both are stored in the system and in the mobile device of the user. If a user wants to access a network, the security system of the network sends a random number to the mobile device. The mobile device encrypts the random number with its security key and algorithm and sends the encrypted number back to the network. The security system of the network encrypts the random number itself and compares the result with the number sent by the mobile device. If both numbers are equal, user authentication succeeded. Using this concept the security key is never sent over the network.

The transmission of data in a digital wireless WAN is encrypted as well. When a connection is established, a random session key is generated. With this session key and a security algorithm, a security key is generated. With this security key and another security algorithm the data is encrypted. A different session key is generated for each connection and as a result a different security key is used for each connection to encrypt the transmitted data.

However, these security mechanisms are not sufficient since, for instance, GSM can easily be intercepted by externals as described in [Pesonen, 1999].

5.   MOBILE AUTHENTICATION

In this section we describe our vision of using mobile devices for authentication purposes.

Authentication is the process of establishing the identity of one party to another [Sandhu&Samarati, 1996]. Authentication verifies the (claimed) *identity* of a user (or a process on behalf of a user) using a particular *identifier* that has to be provided by the user or the process. Well-known types of identities/identifiers are user name/password, certificate/digital signature, or biometric features that do not necessarily require claiming the identity in advance.

We then call authentication to be mobile, if the following criteria hold:
a) The same identity can be used for any service.
b) The same identifier is always available.
c) The same identifier is always applicable.

Following these criteria, authentication based on "what a user knows" (e.g. passwords) is not necessarily mobile, since it is not the same identity that can be used for different services (criterion a). In fact, a user has to have a different account for every service although the account's name could be equal. Furthermore, authentication based on "what a user is" (biometric approaches) is also not necessarily mobile, though the same identifier is always available (criterion b) but not always applicable (criterion c), since biometric interfaces are not standard equipment, yet.

Hence, we specifically concentrate on mobile tokens ("what a user has"), which further allows us to extend basic authentication in order to provide services such as pseudonymous or anonymous authentication (compare section 5.1). These advanced services have to be implemented on the user's device (the token). On the one hand, if biometric authentication was used, a user cannot authenticate using multiple identities in order to foster privacy. On the other hand, even though password based approaches would allow using different pseudonymous the user has to remember all user name / password combinations – an extremely cumbersome process.

Based on mobile tokens, we envision the following scenario for mobile authentication services: we assume a token that is highly portable and personally valuable to the user (e.g. a mobile phone) and a short-range wireless communication technology with ad-hoc properties that is becoming a standard on nearly any kind of mobile and stationary devices (e.g. Infrared, Bluetooth). The mobile phone serves as the token holding the identifier for proving one's identity.

5.1 Degrees of Security

Different applications may require different degrees of security regarding mobile authentication. In the following, we describe simple identification, identification with authentication, pseudonymous authentication, and anonymous authentication.

Mobile identification is a service that provides identification but no authentication. Thus, security requirements are very low. This service is useful to e.g. automatically sign in on attendance lists, which traditionally does not require verifying the identity that someone claims to have.

Mobile authentication is required in cases when a service has to be reasonably sure about the correctness of the identity of the accessing person. For instance, when a computer screen should be unlocked the identity claimed by the user has to be checked.

Mobile pseudonymous authentication can be used in situations when a person wants to identify with a fake name but wants to reuse the same name every time he or she uses the same service. For different services, different identities are claimed to prohibit profiling across services. Other people should not be able to claim the (fake) identity. Therefore authentication is required.

Mobile anonymous authentication allows a user to remain anonymous even if the service requires authentication. For instance, on the Web many services require identification (e.g. username) and some sort of authentication such as a password that is emailed to an existing email account; however, since users want to protect their privacy they will want to generate anonymous accounts automatically.

5.2   Realization Issues

We identified a number of issues when starting pilot applications on mobile authentication. The issues are user/device binding, use of communication protocol features, user intention, multiple devices in place, and un-intentional tracking as explained below.

We assume that a mobile device represents a user based on an a-priori authentication between the device and the user. Mobile devices can then be identified by their device name. Unfortunately, the device name can easily be changed by the user and therefore provides no assurance that the device really is the device it claims to be. However, for simply identifying devices the device name suffices. Furthermore, the identifier used for mobile authentication has to be securely stored on the mobile device. The facilities provided by current mobile devices are however insufficient and the topic has to be further researched.

Depending on the wireless technology used it is feasible to employ security mechanisms provided by the communication protocol itself. For instance, Bluetooth supports three different levels of security. Bluetooth devices can authenticate using a 128bit key – the highest level offered by the protocol itself. Infrared, however, does not support any form of authentication. It is therefore necessary to implement all security features required for authentication on top of the protocol. To abstract the different levels of security intrinsically provided by the different wireless protocols we implement an API that offers two basic services: (1) registering a new user and (2) un-registering users. The API hides the different ways in which the registration is performed from the programmer so that the programmer does not need to be aware of which wireless technology is employed.

The user's intention to authenticate to a particular service may be difficult to recognize, also depending on the underlying communication protocol. Using Infrared, for instance, the approach could be that the user registers by establishing a line-of-sight connection. The next time he establishes a connection again he is unregistered. Connecting yet another time would register again etc. Bluetooth is a protocol that does not require a line-of-sight connection. It is therefore difficult to establish whether a user truly intended an action. For instance, by simply passing by a computer a user does not necessarily want to log in. Therefore, the authentication process has to work differently. One approach could be that as soon as the user enters the area he is authenticated; leaving the area he automatically un-registers. No specific action is required. This, however, may cause a problem because the user gave no explicit consent to authentication when registering. If this consent is required, the mobile device should not automatically register the user but instead display a dialog asking the user whether he wants to authenticate or not.

Knowing what the user really intends to do – by explicitly asking for his consent – is especially important in settings where multiple devices are located in one place. For instance, when a user enters a room he might want to log into a computer but since the location data provided by some technologies (e.g. Bluetooth) is not detailed enough it is impossible to know which computer the user approaches. Moreover, he might also decide not to use a computer at all.

A user carrying a Bluetooth device can be tracked and identified. We envision some method to protect the user's privacy – he should be able to choose which kind of devices or services may observe his presence and which not. Furthermore, there are also services that do not require a username. In these cases no pseudonyms have to be generated and one would assume that by simply establishing a connection the user remains anonymous. However, this is not the case as wireless technologies such as Bluetooth automatically authenticate whenever they establish a connection. Even though the service may not require a login, the user can still be traced because the authentication information is available at the access node. The basic idea to avoid this form of tracking is to regularly change the device address. This prevents the profiling of users that connect to access nodes.

6. CONCLUSIONS AND OUTLOOK

In this paper, we gave an overview of security in mobile communication and mobile device security. Having identified shortcomings we elaborated on the requirements for mobile authentication, gave a definition of mobile authentication and showed which services can be offered based on mobile authentication. Our approach is based on using mobile devices that possess processing capabilities in order to implement services such as mobile authentication and mobile pseudonymous authentication.

The main contribution of this paper is to highlight the rationale for such an approach, to identify the prerequisites, and to argue that a relatively simple API suffices to provide such services. This approach has its advantages over previous forms by relieving the application developer from the burden of having to design, implement, and test authentication mechanisms.

However, two major obstacles were identified that need to be addressed in future efforts in this direction. Firstly, the security of the mobile device itself is not sufficient; this lack of security jeopardizes the reliability of the whole process of authentication. Second, even though Bluetooth is a well-defined standard, integrating it into applications still remains a challenge. We experienced that many program-level APIs do not work reliably with certain devices or some Bluetooth devices do not seem to be supported at all. Moreover, even if a connection can be established it is not sure that this connection can be reestablished. In some cases we had success rates well below 10% without any obvious cause.

Once the aforementioned issues have been addressed, we are convinced that mobile authentication will be useful to implement e.g. mobile signatures scenarios. *Mobile signatures* are digital signatures that are applied to documents displayed on the mobile device. Unlike traditional digital signatures, the mobile device is used as a secure viewer. Therefore the user does not have to trust the external viewer but can be sure that the documents he or she signs are correctly displayed.

REFERENCES

[Bluetooth, 2002]   Bluetooth SIG Home page, http://www.bluetooth.org, last accessed on March 6, 2002.

[Borisov, 2001] Borisov, Goldberg, Wagner, Intercepting Mobile Communications: The Insecurity of 802.11, http://www.isaac.cs.berkely.edu/isaac/mobicom.pdf, last accessed on March 6, 2002.

[Eckert, 2000]   Eckert, Mobile Devices in E-Business – New Opportunities and New Risks, Proceedings of SIS 2000, Zurich, 2000.

[Gosh, 2001]Ghosh, K.A., and Swaminatha, T.M. Software security and privacy risks in mobile e-commerce, Communications of the ACM, Volume 44 (2), Feb 2001, pp 51-57

[Hansmann et al., 2001]   Hansmann, Merk, Nicklous, Stober, Pervasive Computing-Handbook, Springer Verlag, 2001.

[IRDA, 2002]   The Infrared Data Association, http://www.irda.org, accessed March 6, 2002.

[Kelly, 2001]   Kelly, Chair of IEEE 802.11 Responds to WEP Security Flaws, February 15, 2001, http://slashdot.org/articles/01/02/15/1745204.shtml, last accessed on March 6, 2002.

[Microsoft, 2001]   Securing the handheld environment – An enterprise Perspective, White Paper, Microsoft, 2001, http://www.microsoft.com/mobile/enterprise/papers/security.asp (last visited Feb 19, 2002);

[Palm, 2001]   Pocket PC Security, White Paper, Palm, 2001, http://www.palm.com/enterprise/resources/securing/index.html (last visited Feb 19, 2002);

[Palowireless, 2001]   Mc Daid, Bluetooth Security, Parts 1, 2, and 3, http://www.palowireless.com/bluearticle/cc1_security1.asp, http://www.palowireless.com/bluearticle/cc1_security2.asp, http://www.palowireless.com/bluearticle/cc1_security3.asp, last accessed on March 6, 2002.

[Pesonen, 1999] Pesonen L., GSM Interception, Helsinki University of Technology, Dpt. Of Computer Science and Engineering, November 21, 1999, last accessed on March 6, 2002.

[RSA, 2002]   RSA Security Inc., http://www.rsa.com, last accessed on March 6, 2002.

[SANS, 2001]   Robert E. Mahan, Security in Wireless Networks, SANS Institue, http://rr.sans.org/wireless/wireless_net3.php, last visited:March 6, 2002.

[Sandhu&Samarati, 1996] Sandhu R.S., Samarati P. Authentication, Access Control, and Audit. ACM Computing Surveys, Vol. 28, No. 1, March 1996.

[SCCH, 2001]   Gruber, Wolfmaier, State of the Art in Wireless Communication, Technical Report SCCH-TR-0171, Software Competence Center Hagenberg, www.scch.at, 2001.

[Sutherland,   2000],   Sutherland,   Bluetooth   Security: An   Oxymoren?, http://www.mcommercetimes.com/Technology/41, last accessed on March 6, 2002.

[Vainio, 2000]   Vainio J., Bluetooth Security, May 25, 2000, http://www.niksula.cs.hut.fi/~jiitv/bluesec.html, last accessed on March 6, 2002.

[Walke, 2000]   Walke, Mobilfunknetze und ihre Protokolle – Band 1, B. G. Teubner Verlag, Stuttgart, 2000].

CORE: A COLLABORATIVE REPUTATION MECHANISM TO ENFORCE NODE COOPERATION IN MOBILE AD HOC NETWORKS

Pietro Michiardi, Refik Molva

Abstract: Countermeasures for node misbehavior and selfishness are mandatory requirements in MANET. Selfishness that causes lack of node activity cannot be solved by classical security means that aim at verifying the correctness and integrity of an operation. We suggest a generic mechanism based on reputation to enforce cooperation among the nodes of a MANET to prevent selfish behavior. Each network entity keeps track of other entities' collaboration using a technique called reputation. The reputation is calculated based on various types of information on each entity's rate of collaboration. Since there is no incentive for a node to maliciously spread negative information about other nodes, simple denial of service attacks using the collaboration technique itself are prevented. The generic mechanism can be smoothly extended to basic network functions with little impact on existing protocols.

1. INTRODUCTION

A simulation study presented in [1] showed that the performance of MANET severely degrades in face of simple node misbehavior. Unlike networks using dedicated nodes to support basic functions like packet forwarding, routing, and network management, in ad hoc networks, those functions are carried out by all available nodes. This very difference is at the core of the increased sensitivity to node misbehavior in ad hoc networks.

If a priori trust relationship exists between the nodes of an ad hoc network, entity authentication can be sufficient to assure the correct execution of critical network functions. A priori trust can only exist in a few special scenarios like military networks and requires tamper-proof hardware for the implementation of critical functions. Entity authentication in a large network, on the other hand, raises key management requirements.

If tamper-proof hardware and strong authentication infrastructure are not available, the reliability of basic functions like routing can be endangered by any node of an ad hoc network. The correct operation of the network requires not only the correct execution of critical network functions by each participating node but it also requires that each node performs a fair share of the functions. No classical security mechanism can help counter a misbehaving node in this context.

Node misbehavior that affects network operations (routing, packet forwarding) may range from simple selfishness or lack of collaboration due to the need for power saving to active attacks aiming at denial of service (DoS) and subversion of traffic. *Selfish nodes* use the network but do not cooperate, saving battery life for their own communications: they do not intend to directly damage other nodes. *Malicious nodes*, on the other hand, aim at damaging other nodes by causing network outage by partitioning while saving battery life is not a priority.

A basic requirement for keeping the network operational is to enforce ad hoc nodes' contribution to network operations despite the conflicting tendency of each node towards selfishness as motivated by the scarcity of node power. We propose a mechanism called CORE to enforce node cooperation based on a collaborative monitoring technique. CORE is suggested as a generic mechanism that can be integrated with any network function like packet forwarding, route discovery, network management, and location management. Each network entity in CORE keeps track of other entities' collaboration using a technique called reputation. The reputation metric is computed based on data monitored by the local entity and some information provided by other nodes involved in each operation. An interesting feature of the CORE mechanism is that denial of service attacks based on malicious broadcasting of negative ratings for legitimate nodes are prevented.

The remainder of the paper is organized as follows: section 2 introduces the basic reputation concept underlying the CORE mechanism, the generic CORE mechanism presented in section 3 is then illustrated with the applications of this mechanism to packet forwarding and routing functions in section 4.

2. THE REPUTATION CONCEPT

In our scheme, MANET nodes can be thought of as members of a community (or subjects) that share a common resource. The key to solve problems related to node misbehavior derives from the strong binding between the utilization of a common resource and the cooperative behavior of the members of the community. Thus, all members of a community that share resources have to contribute to the community life in order to be entitled to use those resources. However, the members of a community are often unrelated to each other and have no information on one another's behavior. We believe that reputation is a good measure of someone's contribution to common network operations. Indeed, reputation is usually defined as the amount of trust inspired by a particular member of a community in a specific setting or domain of interest. Members that have a good reputation, because they helpfully contribute to the community life, can use the resources while members with a bad reputation, because they refused to cooperate, are gradually excluded from the community.

The approach presented in this section is used as a basis for the security mechanism that solves the problems due to misbehaving nodes by incorporating a reputation mechanism that provides an automatic method for the social mechanisms of reputation. Furthermore the formulae presented in the following sections are conceived in order to minimize problems due to false detection of a nodes' misbehavior. As an example, disadvantaged nodes that are inherently selfish due to their precarious energy conditions shouldn't be excluded from the network using the same basis as for malicious nodes: this is done with an accurate evaluation of the reputation value that takes into account a sporadic misbehavior.

2.1 Definitions

This section presents the three types of reputation used in our scheme and shows how they are combined. Reputation is formed and updated along time through direct observations and through information provided by other members of the community. Furthermore, we take the stance that reputation is compositional: the overall opinion on an entity that belongs to the community is obtained as a result of the combination of different type of evaluations. We define a subjective reputation, an indirect reputation and a functional reputation.

2.1.1   Subjective Reputation

We use the term subjective reputation to talk about the reputation calculated directly from a subject's observation. A subjective reputation at time t from subject s_i point of view is calculated using a weighted mean of the observations' rating factors, giving more relevance to the past observations.

The reason why more relevance is given to past observations is that a sporadic misbehavior in recent observations should have a minimal influence on the evaluation of the final reputation value: as a result, it is possible to avoid false detections due to link breaks and to take into account the possibility of a localized misbehavior caused by disadvantaged nodes.

The general formula to calculate a subjective reputation is:

$$r_{s_i}^t\left(s_j|f\right) = \sum \rho(t,t_k) \cdot \sigma_k$$

where $r_{s_i}^t\left(s_j|f\right)$ stands for the subjective reputation value calculated at time t by subject s_i on subject s_j with respect to the function f.

$\rho(t,t_k)$ is a time dependent function that gives higher relevance to past values of σ_k.

σ_k represents the rating factor given to the *k-th* observation: we use a scale that goes from -1 for a negative impression (meaning that the observed result doesn't match with the expected result) to +1 for a positive impression (i.e. when the observed and the expected results coincides).

When the number or the quality of observations collected since time t are not sufficient, the final value of the subjective reputation takes the 0 value, which is used for a neutral impression.

Finally, given that $\sigma_k \in [-1,1]$ and that $\rho(t,t_k)$ is a normalized value, also $r_{s_i}^t\left(s_j|f\right) \in [-1,1]$.

Note also that the set $\{s_j\}$ is restricted to the set of the neighbors of subject s_i. We use the term neighbor to refer to a subject that is within wireless transmission range of another subject.

2.1.2   Indirect Reputation

In our scheme, the subjective reputation is evaluated only considering the direct interaction between a subject and its neighbors. With the introduction of the indirect reputation measure we add the possibility to reflect in our model a characteristic of complex societies: the final value given to the reputation of a subject is influenced also by information provided by other members of the community.

In the reminder of the paper, $ir_{s_i}^t\left(s_j|f\right)$ denotes the indirect reputation of subject s_j collected by s_i at time t for the function f.

The information collected through indirect reputation can take only positive values: denial of service attacks based on malicious broadcasting of negative ratings for legitimate nodes are thus prevented.

2.1.3   Functional Reputation

We use the term functional reputation to talk about the subjective and indirect reputation calculated with respect to different functions f. With the introduction of this last type of reputation in our model we add the possibility to calculate a global value of a subject's reputation that takes into account different observation/evaluation criteria. As an example, a subject s_i can evaluate the subjective reputation $r_{s_i}^t\left(s_j|f(\text{packet forwarding})\right)$ of subject s_j with respect to the packet forwarding function and the subjective reputation $r_{s_i}^t\left(s_j|f(\text{routing})\right)$ with respect to the routing function and combine them using different weights to obtain a global reputation value on subject s_j.

2.1.4   Combination of reputation information for multiple functions

Reputation information is combined using the following formula:

$$r_{s_i}^t(s_j) = \sum w_k \cdot \left\{ r_{s_i}^t\left(s_j|f_k\right) + ir_{s_i}^t\left(s_j|f_k\right) \right\}$$

where w_k represents the weight associated to the functional reputation value.

$r_{s_i}^t(s_j)$ represents the global reputation value that is evaluated in every node: it is the aggregate reputation definition.

The choice of the weights w_k used to evaluate the global reputation has to be accurate because it can affect the overall system robustness. The simulation study carried out in [1] pointed out that even if the enforcement of the execution of both the packet forwarding function and the routing function are mandatory, the former has an important impact on the global performances compared to the latter. This is why a good choice for w_k would emphasize the correctness of the packet forwarding function when evaluating the overall reputation for a node.

2.1.5   Validation mechanism

Each type of reputation is obtained as a combination of different observations made by a subject over another subject with respect to a defined function f every observation is related to the correct execution of f. It is necessary to define a validation mechanism (based on feed back information) that compares the observed results and the expected results and checks whether they coincides or not. If the objectives have been reached (i.e. observed and expected results coincides) then the rating factor σ_k associated to the *k-th* observation will be positive, while if the observation shows that the expected results are not reached (i.e. the function f has not been correctly executed) then the rating factor will be negative. More details on the validation mechanism will be given in the section 3.1.3 where we consider a possible implementation: the watchdog mechanism.

3. THE CORE SCHEME

This section presents the CORE scheme in details, starting from the definition of the components that participate to the collaborative reputation mechanism and concluding with the description of the complete process in which the different parts are involved.

3.1 Components

3.1.1 Network entity

The network entity corresponds to a mobile node. Each entity s_i is enriched with a set of Reputation Tables (RT) and a watchdog mechanism (WD). The RT and the WD together constitute the basis of the collaborative reputation mechanism presented in this paper. These two components allow each entity to observe and classify each other entity that gets involved in a request/reply process, reflecting the cooperative behavior of the involved parts. The classification of the entities based on their behavior is then used to enforce the strong binding between the cooperative behavior of a subject and the utilization of the common resources made available by all the other entities of the network.

We use the notation *requestor* when referring to a network entity asking for the execution of a function *f* and the notation *provider* when referring to any entity supposed to correctly execute *f*. We also use the notation *trusted entity* when referring to a network entity with a positive value of reputation.

3.1.2 Reputation Table

The Reputation Table (RT) is defined as a data structure stored in each network entity. Each row of the table includes the reputation data pertaining to a node. Each row consists of four entries: the unique identifier of the entity, a collection of recent subjective observations made on that entity's behavior, a list of the recent indirect reputation values provided by other entities and the value of the reputation evaluated for a predefined function. Each network entity has one RT for each function that has to be monitored.

3.1.3 The Watchdog mechanism

The watchdog (WD) mechanism implements the validation phase depicted in section 2.1 and it is used to detect misbehaving nodes. Every time a network entity ($s_{i,m}$, monitoring entity) needs to monitor the correct execution of a function implemented in a neighboring entity ($s_{j,o}$, observed entity), it triggers a WD specific to that function (f). The WD stores the expected result $e_r(f)$ in a temporary buffer in $s_{i,m}$ and verifies if the observed result $o_r(f)$ and $e_r(f)$ match. If the monitored function is executed properly then the WD removes from the buffer the entry corresponding to the $s_{j,o},e_r(f)$ couple and enters in an idle status, waiting for the next function to observe. On the other hand, if the function is not correctly executed or if the couple $s_{j,o},e_r(f)$ remains in the buffer for more than a certain time out, a negative value to the observation rating factor σ_k is reported to the entry corresponding to $s_{j,o}$ in the RT and a new reputation value for that entity is calculated. It should be noticed that the term *expected result* corresponds to the correct execution of the function monitored by the WD, which is substantially different from the final result of the execution of the function.

3.2   Protocol

The CORE scheme involves two types of protocol entities, a *requestor* and one or more *providers*, that are within the wireless transmission range of the *requestor*. The nature of the protocol and the mechanisms on which it relies assure that if a provider refuses to cooperate (i.e. the request is not satisfied), then the CORE scheme will react by decreasing the reputation of the provider, leading to its exclusion if the non-cooperative behavior persists. For sake of simplicity, the following scenarios are related to the execution of the protocol between a *requestor* and one *provider*.

3.2.1   Protocol execution when no misbehavior is detected

First, the *requestor* asks for the execution of a function f to the *provider*. It then activate the WD related to the *provider* for the required f and waits for the outcome of the WD within a predefined time out. Since the two parties correctly behave, the outcome of the WD assures that the requested function was correctly executed and the *requestor* disarms the WD.

We suppose that the reply message corresponding to the result of the execution of function *f* includes a list of all the entities that correctly participated to the protocol: the *requestor* uses this indirect information to update its RT and enters in an idle mode.

3.2.2   Protocol execution when misbehavior is detected

As described in the previous scenario, the *requestor* asks for the execution of a function *f* and arms the related WD, waiting for the outcome. Since we suppose that the provider does not cooperate, the outcome of the watchdog will be negative. The *requestor* will then update the entry in the RT corresponding to the misbehaving entity with a negative factor and will enter in an idle mode.

3.2.3   Request made by a misbehaving entity

We describe here the process that any entity receiving a request has to follow. Upon receiving the request for the execution of a function *f* the entity checks the reputation value evaluated for the *requestor* in its global RT. If the reputation value is negative then the entity will not execute the requested function. It has then the choice whether to notify or not the denial of service. A detailed analysis on the best practice will be presented in section 3.4.

3.3   RT updates and distribution

We focus now on the mechanism used to update and distribute reputation information. RTs are updated in two different situations: during the request phase of the protocol and during the reply phase corresponding to the result of the execution of *f*.

In the first case, it is possible to notice that only the subjective reputation value is updated. If the outcome of the WD shows that the *provider* did not cooperate, a negative rating factor will be assigned to the observation and consequently the reputation related to the misbehaving entity will decrease. If no misbehavior is detected, the RTs are not updated.

In the second case, only the indirect reputation value is updated. We suppose that the reply message contains a list of all the entities that correctly behaved: the indirect reputation will be positive and consequently the reputation related to the cooperating entities will increase.

The reason why only positive rating factors can be distributed among the entities while the negative rating factors are evaluated locally derives from a possible attack to the protocol. If negative factors could be spread around, it would be simple for a misbehaving entity to distribute false information about other entities in order to initiate a denial of service (DoS) attack. The protocol presented in this paper allows only the distribution of positive rating factors: if we suppose a scenario where collusion between misbehaving entities is impossible, then there would be no advantage for a misbehaving entity to distribute positive rating factors to other unknown entities. Furthermore, reputation information is distributed and updated only during the reply phase avoiding a indiscriminate broadcast of bogus information.

Reputation values calculated for each entry of the RT are not constant: if the reputation value is positive then it is decremented along time. The reason why we decided to decrement positive reputation values comes from a possible attack to the CORE scheme: if a network entity enters in an idle status for most of the time except when it has to communicate, its reputation has to be decreased, even if during the active time it cooperates to the network operation.

3.4   Cooperation Enforcement

This section describes how reputation information is used to enforce cooperation between entities. Reputation is directly related to the cooperative behavior of an entity: if the reputation value is negative then the entity is classified as a misbehaving entity while if the reputation value is positive then the entity is tagged as a trusted entity. The execution of a function requested by any *requestor* is conditioned by the corresponding reputation value stored in the global RT of the *provider*: when this reputation value is negative then the provider will deny the execution of the requested operation.

There is no advantage for an entity to misbehave because any resource utilization will be forbidden. Reputation is hard to build because positive rating factors are acquired only in the reply message which contains the list of all the network entities that cooperated to obtain of the final result of the requested function. On the other hand, negative rating factors are attributed every time the outcome of the WD is negative. Even if reputation is not linearly decreased for every negative rating factor in order to avoid false evaluations (e.g. apparent misbehavior due to link breaks), a persistent non-cooperative behavior compromises normal resource utilization leading to the exclusion of the misbehaving entity from the network.

4. APPLICATIONS

4.1 Background and assumptions

This section outlines the assumptions that were made regarding the properties of the physical and network layer of the MANET. Throughout this paper we assume bi-directional communication symmetry on every link between the nodes. Furthermore the routing protocol that has been used as a basis for the study of the CORE scheme is the Dynamic Source Routing (DSR) protocol. In addition, we assume wireless interfaces that support promiscuous mode operation. The watchdog technique presented in section 3.1.3 relies on the promiscuous mode operation and has some weaknesses that have been presented in [2].

4.2 Node misbehavior model

The node misbehavior model used in this paper take inspiration from the threats presented in [1]. The research presented in [1] pointed out two types of misbehavior: a selfish behavior and malicious behavior. The protocol presented in this paper focuses on the node selfishness problem.

4.3 Application of CORE to the DSR Route Discocery function

Route discovery allows any node in the ad hoc network to dynamically discover a route to any node in the ad hoc network, whether directly reachable within wireless transmission range or reachable through one or more intermediate network hops through other nodes. A node initiating a route discovery broadcasts a route *request* message which may be received by those nodes within wireless transmission range of it. When any node receives a route request message it processes the request and if the target of the request is unknown it appends the node's own address to the route record in the route request packet and re-broadcast the request. If the route discovery is successful the initiating node receives a route *reply* message listing a sequence of network hops through which it may reach the target.

As described in section 3.2, the CORE scheme involves a *requestor* and one or more *providers* that are within the wireless transmission range of the *requestor*. The CORE protocol can be thought of as a layer on top of the DSR protocol, and the function f that has to be monitored corresponds to the Route Discovery function of the DSR protocol. The WD mechanism is able to detect any misbehaving node that does not participate to the Route

Discovery phase of the protocol and the evaluation of the reputation value reflects any node misbehavior. Node misbehavior is detected in the request phase of the Route Discovery function while the reply phase informs the initiator and the intermediate nodes on the identity of the network entities that participated to the Route Discovery phase: reputation value is updated to reflect the positive rating factors assigned to the cooperating nodes.

Every node stores a set of RTs that are used to classify other nodes of the network: route requests originating from nodes classified as cooperating entities will be served properly whereas routing service will be denied to route requests issued by misbehaving nodes. Only a cooperative behavior allows an entity to change its reputation value from negative to positive: nodes are stimulated to participate to the Route Discovery function if they want to be served when they need to communicate.

4.4 The CORE scheme applied to the Packet Forwarding function

Similarly, the CORE scheme can be used to monitor the Packet Forwarding (PF) function. Once a node has obtained a valid route to the destination through the DSR Route Discovery function, it can start sending data packet to its target. Each network entity belonging to the path from the source to the destination has to perform the PF function in order transfer the data packets. The WD mechanism can be used to detect any misbehaving nodes that refuse to cooperate to the PF and the evaluation of the reputation value reflects any node misbehavior.

As opposed to the Route Discovery function, the PF function does not offer separate operations that can be qualified as request and reply phases. However, if an acknowledgment (ACK) packet can be included in the original data transfer protocol for the purpose of security, the transfer of the data packet can be thought of as the request phase while the transfer of ACK can be considered as the reply phase.

As described in section 4.3 any node misbehavior is detected in the request phase of the PF function while the reply phase informs the initiator and the intermediate nodes on the identity of the network entities that participated to the PF: reputation value is updated to reflect the positive rating factors assigned to the cooperating nodes.

Every node stores a set of RTs that are used to classify other nodes of the network with respect to the PF function. The execution of the PF function is granted for any node classified as a cooperating entity while it is denied for misbehaving nodes. Only a cooperative behavior allows an entity to change its reputation value from negative to positive: nodes are stimulated to participate to the PF function if they want their own data packet to be forwarded to the destination.

5.   RELATED WORK

The area of ad hoc networking has been receiving increasing attention among researchers in recent years and a variety of routing protocols targeted specifically at the ad hoc networking environment have been proposed. However, very few researchers focus on the selfishness problem in MANET and existing work in this area is still in its infancy.

In [2], the authors consider the case in which some misbehaving nodes agree to forward packets but fail to do so. In order to solve this problem, they propose two mechanisms: a watchdog, in charge of identifying the misbehaving nodes, and a pathrater, in charge of defining the best route circumventing these nodes. The paper shows that these two mechanisms make it possible to maintain the total throughput of the network at an acceptable level, even in the presence of a high amount of misbehaving nodes (e.g., 40%). However, the selfishness of the nodes does not seem to be castigated; on the contrary, by the combination of the watchdog and the pathrater, the misbehaving nodes will not be bothered by the transit traffic, while still enjoying the possibility to generate and to receive traffic.

CORE differs from the watchdog-pathrater scheme as follows:

- in CORE misbehaving nodes are stimulated to contribute to the network operations in order to be able to use network services, the pathrater mechanism helps a legitimate user to avoid using misbehaving nodes;

- CORE is a generic mechanism that can be integrated with several network and application layer functions whereas the watchdog-pathrater scheme is specifically designed for routing;

- unlike the pathrater technique the reputation mechanism in CORE does not allow a node to distribute negative ratings about other nodes, so unlike the pathrater technique, CORE can resist to simple denial of service attacks that use the security mechanism itself.

In [7], the authors present two important issues targeted specifically at the ad hoc networking environment: first, end-users must be given some incentive to cooperate to the network operation (especially to relay packets belonging to other nodes); second, end-users must be discouraged from overloading the network. The solution presented in their paper consists in the introduction of a virtual currency (that they call Nuglets) used in every transaction. Two different models are described: the Packet Purse Model and the Packet Trade Model. In the Packet Purse Model each packet is loaded with nuglets by the source and each forwarding host takes out nuglets for its forwarding service. The advantage of this approach is that it discourages users from flooding the network but the drawback is that the source needs to know exactly how many nuglets it has to include in the packet it sends. In the Packet Trade Model each packet is traded for nuglets by the intermediate nodes: each intermediate node buys the packet from the previous node on the path. Thus, the destination has to pay for the packet. The direct advantage of this approach is that the source does not need to know how many nuglets need to be loaded into the packet. On the other hand, since the packet generation is not charged, malicious flooding of the network cannot be prevented. There are some further issues that have to be solved: concerning the Packet Purse Model, the intermediate nodes are able to take out more nuglets than they are supposed to; concerning the Packet Trade Model, the intermediate nodes are able to deny the forwarding service after taking out nuglets from a packet.

6.   FUTURE WORK

The security approach presented in this paper will be completed with an accurate analysis and classification of denial of service attacks specific to the ad hoc networks environment. Indeed, in this paper we considered only selfishness as a specific issue to address: selfish nodes, however, do not intend to directly damage other nodes while the misbehavior is due to their need to save battery life for their own communications. Our ongoing research is evaluating the robustness of the proposed scheme when we consider also malicious nodes that aim at damaging other nodes. In this case, active denial of service attacks can be performed by malicious nodes and our work focus on the definition of other possible attacks.

Furthermore, we focus also on the definition of a formal method, based on the game theory, to analytically prove the robustness of our scheme: we expect to demonstrate that the security mechanism exposed in the paper is compliant to our security objectives.

An in-depth analysis of our security scheme is ongoing using our simulation environment. Our goal is to implement a wide choice of attacks using the QualNet network simulator: we enhanced our software by adding passive denial of service attacks perpetrated on the packet forwarding function and the routing function and we plan to add new features including active denial of service attacks and traffic subversion. We also aim at extending our misbehavior model in order to consider eventual collusions between malicious entities.

The analysis of the simulation results is based on an appropriate metric we defined in order to give emphasis to the robustness of a generic security scheme with respect to the percentage of misbehaving nodes present in the network. We also plan to analyze the performances of our mechanism with respect to node mobility and node density: we believe that network characteristics can be used as trigger signals for the fine tuning of our scheme.

7.   CONCLUSIONS

The area of ad hoc network security has been receiving increasing attention among researchers in recent years. However, little has been done so far in terms of the definition of security needs specific to different types of scenario that can be defined for ad hoc networks. We introduced a fundamental distinction between ad hoc networks where an a priori trust relationship exists between the nodes, provided as an example by a common authority, and ad hoc networks where there is no shared a priori trust between the mobile nodes.

Our research is focused on MANET where there is a lack of a priori trust relationship between mobile nodes. Countermeasures against node misbehavior in general and denial of service attacks in particular is our very first concern. In this paper we suggested a generic mechanism based on reputation to enforce cooperation among the nodes of a MANET and to prevent passive denial of service attacks due to node selfishness. This mechanism can be smoothly extended to basic network functions with little impact on existing protocols.

REFERENCES

[1] P. Michiardi, R. Molva. Simulation-based Analysis of Security Exposures in Mobile Ad Hoc Networks. European Wireless Conference, 2002.

[2] S. Marti, T. Giuli, K. Lai, and M. Baker. *Mitigating routing misbehavior in mobile ad hoc networks*. In Proceedings of MOBICOM, 2000.

[3] The Terminodes Project. www.terminodes.org.

[4] L. Blazevic, L. Buttyan, S. Capkun, S. Giordano, J-P. Hubaux, and J-Y. Le Boudec. *Self-organization in mobile ad hoc networks: The approach of Terminodes.* IEEE Communications Magazine, June 2001.

[5] L. Buttyan and J-P. Hubaux. *Enforcing service availability in mobile ad hoc networks.* In proceedings of MobiHOC, 2000.

[6] J.-P. Hubaux, T. Gross, J.-Y. Le Boudec, and M. Vetterli. *Toward self-organized mobile ad hoc networks: The Terminodes Project.* IEEE Communications Magazine, January 2001.

[7] L. Buttyan and J.-P. Hubaux. *Nuglets: a virtual currency to stimulate cooperation in self-organized ad hoc networks.* Technical Report DSC/2001/001, Swiss Federal Institute of Technology -- Lausanne, 2001.

[8] L. Zhou and Z. Haas. *Securing ad hoc networks.* IEEE Network, 13(6):24--30, November/December 1999.

[9] G. Zacharia. Collaborative Reputation Mechanisms for online communities. Master's thesis, MIT, September 1999.

ENABLING ADAPTIVE AND SECURE EXTRANETS

Yves Roudier[1], Olivier Fouache[1], Pierre Vannel[2], and Refik Molva[1]

[1] {roudier,fouache,molva@eurecom.edu}
   Institut Eurécom, 2229 route des Crêtes, B.P. 193, 06904 Sophia-Antipolis, France
[2] {pierre.vannel@gemplus.com}
   Gemplus Labs Labs, Parc d'activités de Gémenos, B.P.100, 13881 Gémenos Cedex, France

Abstract: Extranets are tools that enable an organization to share part of its information system and infrastructure with other parties. Reaching this goal requires shielding from intruders while at the same time dynamically opening intranet resources. This article discusses how should such an extranet be designed. A solution that automates access control definition and enforcement is presented, which also addresses wide scale user management using a capability-based model. A prototype using the SPKI infrastructure is described that offers strong authentication thanks to smart cards.

Key words: extranets, firewalls, authentication, SPKI, Handle System, smart cards

1. INTRODUCTION

Today's business-to-business electronic commerce environments exhibit a major trait: information has to be shared with other parties such as suppliers, customers, partners, etc. On the other hand, totally opening the intranet is not acceptable: information is one of the main assets of a company, and thus needs to be thoroughly protected. Extranets are tools that enable a safe sharing. They interconnect the intranet of an organization with that of another party so that the organization can make its documents, services, computers, etc. available to a partner.

However, even though access control is becoming critical in a corporate extranet, the scale at which it must be managed does not make it easy for, say, two large organizations to interconnect. The number of available resources increases tremendously, as does the potential number of end-user accessing them: this calls for the use of a fine-grained access control to resources. In addition to concerns about the granularity of access control, business relationships have to be implemented more frequently, sometimes almost immediately, and often for short periods of time: information rollout time has become a key factor for businesses. Finally, an intrusion in the corporate network is possible, be it from a hacker on the Internet or from an end-user of the partner organization.

The major problem with these requirements is that administrators cannot be expected to handle extranet security configuration at the frequency that is needed. In order to enable wide scale information sharing with other partners without compromising security, the security devices available in an extranet have to be automatically configured and operated, and in the most transparent way for end-users.

Is it possible to evolve the existing extranet paradigm to make it more suitable for B2B applications? This article proposes the *adaptive extranet paradigm* as a solution to these needs. In such a system, it is the automation and association of several traditional security tools like firewalls, public key certificate infrastructure, smart cards, and intrusion detection systems that enables a strong authentication of users and of their exchanges on a wide scale. This article also describes a prototype that was implemented along these guidelines in the SEVA project [SEVA].

Section 2 motivates the necessity of a new extranet paradigm based on the shortcomings of available technologies for setting up an intranet. Section 3 describes what should an adaptive extranet be made of. Section 4 details the definition and dissemination of application access rights. Finally, Section 5 explains the architecture chosen for enforcing these rights and gives some details about our prototype implementation.

2. TECHNOLOGIES FOR EXTRANETS

Corporate extranets have to provide access to a company resources. Although peer-to-peer technologies offer an attractive way to implement such an access [S-Peer], corporate networks are not ready yet. Many resources are closely tied with the server and service technology with which they were developed. This section considers several approaches to secure access to such resources.

2.1   Extranets vs. Virtual Private Networks

When it comes to securely interconnecting networks, the prevalent network architecture is that of virtual private networks. With this architecture, a company's network can be extended beyond the boundaries of a firewall-controlled area. A virtual private network provides a network-level cryptographic traffic protection between two intranets. A good example of such setup is given by the IP Security Protocol [KeAt98a][KeAt98b], for instance in its tunnel mode. However, the VPN architecture implicitly defines the network as that of a single party only.

On the contrary, extranets are networks in which several partners share their resources and collaborate, as opposed to the VPN single party approach. In contrast with the VPN, which simply provides a tunnel for traffic, extranets must be defined from scratch in terms of the services offered. In particular, this implies defining an agreement on how these services can be used.

2.2   Authentication and Access Control with Firewalls

A first problem of the firewall architecture is that current security architectures are static: a human administrator is constantly required in order to update the filtering performed. Moreover, security entirely depends on the availability of this operator who must shut down any access of a malicious user. Finally this process is error-prone for filtering rules are complex to write.

Dynamic or adaptive firewalls are an attempt at solving this problem. For instance, the SunScreen firewall [SUN] introduces the so-called time of day rules, which are rules activated at a programmed hour. Other firewalls can be connected with an intrusion detection system. These firewalls can close a connection in case of an intrusion, but they do not solve the need for a dynamic configuration of extranets.

Current corporate network architectures generally associate firewall and server authentications: firewalls are first used to filter the traffic based on network-level elements, then users are authenticated on application servers with access control lists. LDAP databases help centralize these ACLs and share information with the firewall. In such an architecture, firewalls do not establish the corporate security perimeter anymore, but simply act as static malicious traffic filters. Application-level information is unfortunately unavailable to the firewall in order to adapt its filtering.

2.3   SOCKS

SOCKS [LGL96] is an IETF approved standard networking proxy protocol that enables hosts to access servers without requiring direct IP reachability. Authentication is possible and SOCKS might appear as a good solution to authentication on a firewall or more generally at the border of an intranet.

However, SOCKS requires modifying every communication call that is made in an client application. This is simply not possible with applications for which the source code is not available, that is the vast majority of applications in a corporate network. A new technology [eBorder] has been introduced recently that solves this problem. However, SOCKS authentication features are rather rudimentary compared to the needs of an extranet.

3.   TOWARDS AN ADAPTIVE EXTRANET ARCHITECTURE

Local area networks, then intranets, still made it possible to identify a person in an organization and grant him one, sometimes several privileges for managing a file system or accessing an application server. The number of users envisioned in extranet architectures radically changed this point of view.

The use of capabilities, materialized by certificates is central here: no other solution makes it possible to manage inter-domain exchanges, that is, authenticating users from several companies dispersed over the Internet. The firewall configuration can be driven by the resolution of these capabilities.

We will call an extranet incorporating such mechanisms an adaptive extranet. The overall architecture of an adaptive extranet can be divided into three tiers, each of which has to handle a part of the complexity of securing business-to-business operations: the service management tier, the user management tier, and the extranet administration tier for configuring all security components.

3.1   Service Management Tier

An extranet consists of services offered by an organization to its partners. Establishing an extranet means deciding, at a high level, what services are available to a partner and being able to update this information. With the sheer volume of information on networks is exponentially growing, it appears that access control to the computers where this information is stored is not sufficient. The granularity of the access control should be smaller, but the multiform nature of digital documents (e.g. web pages, multimedia documents, applications...) as well as their organization make it impossible in practice to have a universal access control scheme.

Digital document references may provide an answer. The Corporation for National Research Initiatives (CNRI) has proposed the Handle System as "a general-purpose global name service enabling secure name resolution over the Internet" [SRL01]. This proposal is currently under work at the IETF. It introduces a unique global namespace for digital resources over the Internet, with a root naming authority and sub-naming authorities, referencing all sub-naming authorities worldwide. An identifier to a digital resource is called a handle and can be resolved into a resource, for example an HTTP URL, or an FTP address.

Handles permit the resource administrator not only to define a uniform reference for accessing a resource, but more importantly, to introduce a very fine granularity of access control over the resources he provides. Access rights can be granted on each handle, independently from the application server and without modifying it. We view the handle system as an extensible framework for storing access control information, which makes it very suitable for defining multiparty application access rights.

The actual architecture of the Handle System, with a unique and public namespace, is not satisfying in the context of an adaptive extranet. The Handle System was thus modified in SEVA in order to introduce a private namespace per extranet. One extranet participant runs the extranet central authority to reference all naming authorities of the other partners (new feature). Every extranet participant runs a Local Handle Server (LHS). A handle on an extranet resource is only visible by extranet participants, an end-user being authenticated thanks to her smart card (new feature).

3.2 User Management Tier

Basically, the ultimate role of an authentication and access control architecture is to define if a user has been granted some right. This can be handled without too many hassles within a company's intranet. However, the several thousands of users from another company are managed by a different administrator than the one in charge of the resources available in the intranet accessed. Furthermore, managing an extranet is more complex: employees join and leave partner companies, partners join the extranet, etc. The successful deployment of any multiparty resource sharing architecture would imply that an important number of users would access a resource, even for a short moment: a capability-based access control model must be used in order to manage users' rights.

In addition, managing the users of another company directly would have privacy implications that a business would preferably avoid. Roles [SaSa97] represent a solution and can be introduced into the capabilities referred to above. In that respect, group certificates make the Simple Public Key Infrastructure (SPKI) ideal for managing the access rights granted over available resources of the extranet.

SPKI [EFL98a][EFL98b][EFL99] is destined to answer access control problems in wide-scale networks and is standardized by IETF: access control capabilities are encoded with authorization certificates. As opposed to X.509 [ITU88], SPKI is not designed for naming a party, but for stating its access rights. SPKI also permits the transitive delegation of access rights, which is essential for spanning the different jurisdictions of an extranet.

Handles offer an adapted resource designation for defining access rights and can be used straightforwardly in SPKI certificates. Handles can also make certificates smaller, for instance, they can reference the different access modes to a resource, which needs not be included in the certificate itself. Finally, handles make it simpler to lookup all the certificates required for proving that the requesting client has access to a resource.

3.3   Extranet Administration Tier

This tier lies in the middle of the two previous tiers and enforces the access control operations based on the information provided by those two tiers. In particular, it has to manage the firewall, and the user logs.

The administration tier also verifies that users have the right to access a resource. Adaptive extranets address applications where people from different corporate networks are accessing resources from another network and working together across the Internet. It thus becomes mandatory to design an architecture that not only protects from Internet hackers' attacks but also from malicious users even though their traffic is supposedly originating from a business partner. Users' rights are thus checked at the corporate system's border, as transparently as possible, then the behavior of authenticated users can be logged by the intrusion detection system. Suspect behaviors can entail the suppression of the user access to the local intranet.

The services available to every partner are also defined in an inital agreement that establishes the extranet. This agreement is used to generate the access rights of the allowed parties, and to configure firewalls and administration stations. XML [W3C00], or more precisely tpaML [tpaML] was selected to express such a contract: it can be used to automate the processing of the agreement that is updated each time a member joins or leaves the extranet. The rules regulating the extranet, like liabilities, security parameters, etc. can also evolve as well as the services offered and thus require establishing a new agreement.

4. ACCESS CONTROL DEFINITION AND MANAGEMENT

Access control definition is tremendously important in the adaptive extranet architecture. Access rights have to be distributed and stored first but they must also be managed. Access right management can be determined from the start thanks to an authorization certificate lifetime for instance, or might be needed suddenly in case of a network intrusion.

4.1 End-User Accreditation

The objective of the accreditation is to transcribe the extranet agreement within a participant's organization: that means deploying the extranet services for the authorized end-users and distributing them the corresponding access rights. Current organizations are hierarchical. The allocation of new services and their associated access rights follows a hierarchical path.

For instance, administrator stations rely on agents for automatically generating new SPKI certificates based on those that are regularly issued for every partner administrator. SPKI Delegation is heavily used in this process. At any hierarchical level, empowered end-users can delegate or restrict the access rights of their subordinates.

SPKI Certificates are very handy for avoiding revocation problems since they become automatically useless after their validity, which can be decided by the administrator, is over. The best way to exploit this feature is to manage certificate issuance through an automated infrastructure: such a system can be envisioned as a set of intelligent agents issuing a new certificate for every user according to the administrator settings. This part of our prototype is still under development based on the JADE agent platform [BPR00].

4.2 The Smart Card: a Management Tool for Access Rights

During the accreditation process, the end-user's smart card plays a key role. This smart card is a portable and personal wallet to access to the extranet services. It can be used to securely store cryptographic keys as well as the reference to services or resources. The smart cards used are Java Cards and offer multi-application support. They consist of a cryptographic card applet and of an embedded secure LDAP-like server [Mac00] where the service or resource references can be stored as well as access rights. The cryptographic card applet is a part of the authentication and access control scheme described in the next section.

The memory of current smart cards is still limited: only a few tens of kilobytes to store programs and data. The smart card is adequate to store secret cryptographic keys and to protect the cardholder's privacy. We thus chose to store only the reference to the certificate: the real certificate is stored into an LDAP server inside the intranet of the user's company rather than directly on the smart card.

4.3   Intrusion Detection and Access Right Management

The authentication of a user does not preclude the possibility of an attack on his part. The company offering the resource might in that case want to revoke the rights of a user performing repeated illicit or suspect operations.

Mobile users introduce yet another threat: these users can access services from a partner of their company or using a laptop computer from an Internet Service Provider, but they reside in a hostile environment. An attacker can more easily hijack their machine with a Trojan horse and then create havoc in the network of one extranet partner.

In both cases, a network intrusion detection system becomes an important part of the access control management infrastructure and these two elements should be interconnected. The snort [Roe] IDS has been integrated within our prototype adaptive extranet using the standardized IDMEF [CuDe01] exchange language. If an intrusion is detected, the liability of a user can be established with the help of the company who handles his identity and rights.

5.   AUTHENTICATION IN THE EXTRANET

Once the access rights have been established by distributing certificates, actual traffic can be exchanged between the two intranets. Access control has to be enforced on this traffic to make it compliant with the access rights previously determined. An adaptive extranet must perform a strong authentication of the traffic exchanged, and in relation with the resources accessed.

5.1   An Authenticating Firewall

Access control must not only be described, but also enforced in conjunction with authentication. Firewalls [ChBe94] provide the most widespread technique for doing so and securing intranets: they enforce access control on the traffic by filtering packets and connections. This filtering security model is attractive because it is performed at intermediate network elements and thus leaves application servers unmodified. This model is also quite easy to deploy, since all filtering operations can be performed in a centralized manner, for all servers of a domain for instance.

However, the most commonly used filtering model is that of the access control list, in which the operations authorized are listed for each user name. The architecture of an extranet requires filtering access based on capabilities instead as explained in Section 3.

5.2   Certificate Based Authentication and Access Control

Preserving performance is an important design guideline. This is why access control is enforced in two steps: application rights of a user are proven by sending a set of SPKI certificates; all traffic from that user is subsequently authenticated using lighter cryptographic operations.

As illustrated in Figure 1, whenever a handle (1) has to be resolved into an application resource (a URL in the example), the initial authentication is performed together with access control resolution (message (3)).

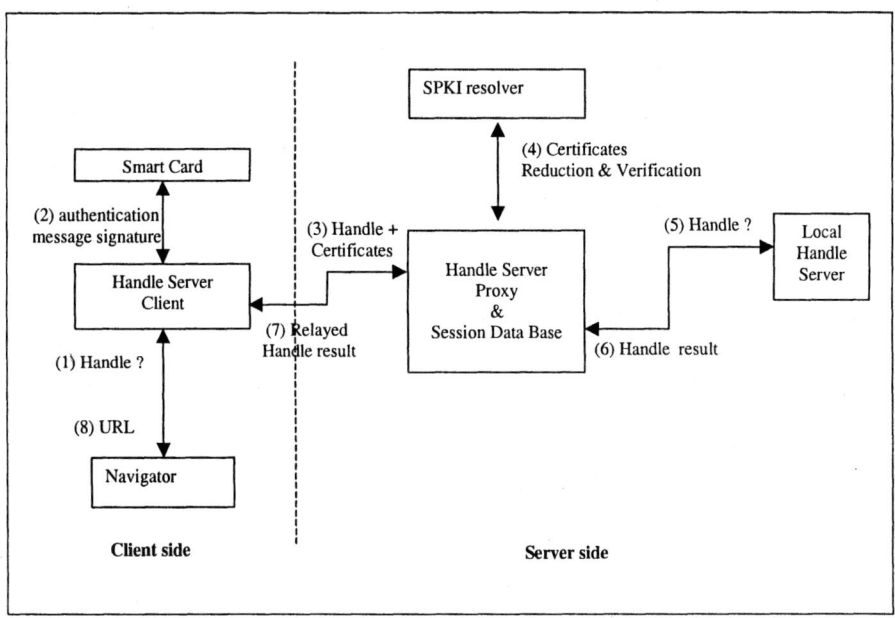

Figure 1. Verification of Access Rights

In our prototype, the Handle Server proxy was added to the set of proxies available in the TIS firewall toolkit [TIS]. It performs the associated SPKI certificate resolution (4) for proving the access rights of the end-user. It then also verifies that the message providing the handle is correctly signed by the smart card, which authenticates the user.

If the user is authenticated, the Handle Server proxy also performs the handle resolution (5) with a handle server modified to run without contacting the handle system root server hosted at CNRI. After this authentication, the user is logged in a session database and can subsequently be identified thanks to a session key transmitted in the authentication message (see section 5.3).

5.3   Traffic Tagging

The purpose of traffic tagging is to ensure that the packets transmitted were sent by the user, the sole entity owning the secret key, without having to pay the price of signing each and every packet transmitted. This has two important consequences: the traffic is strongly authenticated; a user can be logged and identified via his public key, although his name remains unknown to the firewall logging his accesses.

Tagging can be performed through the introduction of a specific communication layer (see Figure 2) in the client workstation. The data transmitted are encapsulated and marked with a cryptographic ticket (3). The ticket establishes the identity of the user based on the session key and ensures at the same time the integrity of the data: it simply consists of the keyed hashing of the data, using a session key established beforehand with the smart card. Encryption has not been provided but might as well be performed between intranets using the session key. However, it is not required for authentication only and we tried to limit the performance impact of the extranet security mechanisms.

A modified socket library intercepting any communication directed towards another SEVA intranet was programmed on Windows and experimented first (it offers a functionality similar to [eBorder]). The TCP packet was thus encapsulated with a specific traffic format, comprising the data, the authenticating cryptographic ticket, and some additional information like the destination address and port. With this network-level approach, it is possible to finely tune the granularity of authentication and thus the buffering of packets.

We also experimented with another technique: we focused on HTTP traffic, and finally implemented tagging via a simple HTTP proxy for its ease of deployment on several platforms (it currently runs on Linux or Windows). Using a proxy has the advantage of making it possible to use application-level information. For instance, instead of accessing resources through a handle, the user just types a URL in his browser and our prototype communication layer is able to convert this URL into one of the handles accessible to the user. Integration with applications becomes totally seamless.

The ticket verification is performed on the firewall by the corresponding proxy (4), in relation with a previously authenticated handle resolution. This proxy verifies that the ticket was constructed correctly. The proxy also decapsulates the original traffic before relaying the connection (5).

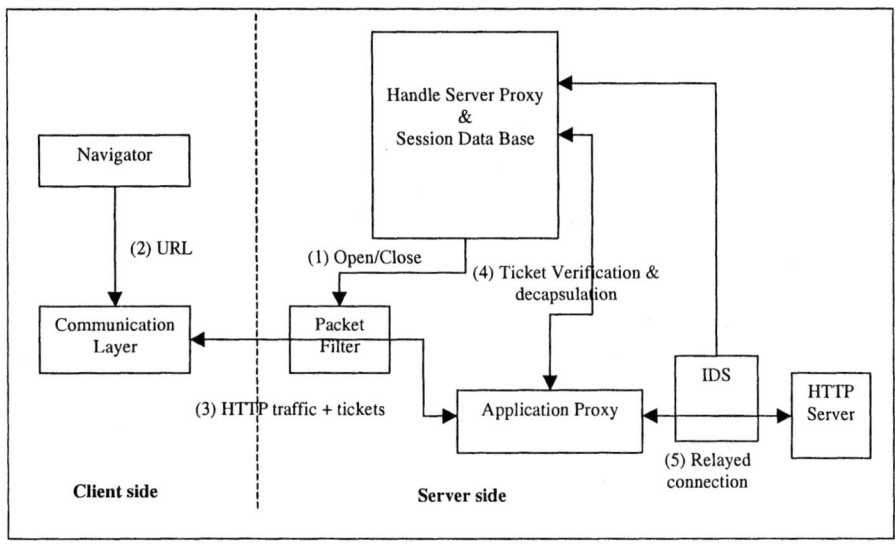

Figure 2. Traffic Tagging

The smart card is used here as a real-world key to the adaptive extranet: it keeps the private key of the user that does not reside on the workstation itself and is used as an essential element in a zero-knowledge authentication protocol. No traffic with another partner can take place on the extranet without the smart card because the ticketing process cannot run short of a valid session key. Only the smart card can generate a session key for authenticating the traffic. Inserting a card in a workstation reader enables the workstation to connect during a fixed time slot. At the end of this slot, a new session key is automatically reestablished with the handle service proxy. For performance reasons however, ticketing cannot performed on the card, which can only exchange protocol data units with a serial line.

6. CONCLUSION

Business-to-business corporate services require the secure interconnection of the networks of different parties. This interconnection is difficult because on one hand, services and the resources that they reference are complex, and on the other hand, identity and end-user's rights can only be defined and handled within the user's company. Furthermore, B2B relations necessitate a very swift configuration of security devices whose frequency cannot be handled by human operators. The security devices used in extranets presently do not address these requirements. Adding security devices like an intrusion detection system to an extranet will not solve these problems *per se* either.

This article proposed to have a certain number of otherwise classical security devices collaborate in a coordinated fashion to extend the capabilities of a corporate firewall in order to realize what can be called an adaptive extranet architecture. Users and access rights are central to such an adaptive extranet and must be handled separately: both can be managed using a SPKI public key infrastructure and a generic notion of resource named a handle. The user's traffic is authenticated between his workstation and the firewall of the corporate network accessed with a lightweight cryptographic scheme so as not to degrade access performance. The tight integration of SPKI authentication with the traffic ticketing process makes it possible to achieve the automatic configuration of the firewall. Smart cards play an important role for authentication as well and are used instead of a password. They are personal security devices that integrate perfectly with the automatic configuration of the extranet security devices thanks to the SPKI model: they enable users to prove his identity or administrators to issue authorization certificates. Finally, even though strong authentication of end-users is enforced, the system does not threaten end-user privacy.

An adaptive extranet prototype has been developed and deployed by the SEVA project team that already implements all security features for access control and certificate automated distribution. Results thus far are very encouraging: the access to a web server can be totally controlled without servers being modified at all by the deployment of this system. The SEVA prototype implementation also demonstrates that client applications can be retrofitted with strong authentication without any modification. We experimented with the introduction of access control for UDP traffic as well, and would be interested to support it more completely in the future.

ACKNOWLEDGEMENT

SEVA is a collaborative project of ATOS, Electricité de France, Eurécom, and Gemplus, and is supported by the French Ministry of Economy, Finances, and Industry and the Réseau National de Recherche en Télécommunications (RNRT).

REFERENCES

[BPR00] F. Bellifemine, A. Poggi, G. Rimassa, and P. Turci. *An Object Oriented Framework to Realize Agent Systems.* in Proc. of WOA 2000 Workshop, Parma, May 2000, pp. 52-57.

[ChBe94] B. Cheswick and S. Bellovin. *Firewalls and Internet Security: Repelling the Wily Hacker.* Addison Wesley, 1994, ISBN 0-201-63357-4

[CuDe01] David Curry, Hervé Debar. *Intrusion Detection Message Exchange Format Data Model and Extensible Markup Language (XML) Document Type Definition* - draft-ietf-idwg-idmef-xml-06.txt. December 2001

[eBorder] http://www.permeotechnologies.com/technology/wpapers.htm. Permeo Technologies. *e-Border white papers.*

[EFL98a] C. M. Ellison, B. Frantz, B. Lampson, R. Rivest, B. M. Thomas, and T. Ylönen. *Simple Public Key Certificate*, Internet draft <draft-ietf-spki-cert-structure-05.txt>, March 1998.

[EFL98b] C. M. Ellison, B. Frantz, B. Lampson, R. Rivest, B. M. Thomas, and T. Ylönen. *SPKI Examples*, Internet draft <draft-ietf-spki-cert-examples-01.txt>, March 1998.

[EFL99] C. M. Ellison, B. Frantz, B. Lampson, R. Rivest, B. M. Thomas, and T. Ylönen. *SPKI Certificate Theory*, RFC 2693, September 1999.

[ITU88] ITU-T. *Recommendation X.509: The Directory - Authentication Framework*, 1988.

[KeAt98a] S. Kent, R. Atkinson. *IP Authentication Header (RFC 2402).* November 1998.

[KeAt98b] S. Kent, R. Atkinson. *IP Encapsulating Security Payload (ESP) (RFC 2406).* November 1998.

[LGL96] M. Leech, M. Ganis, Y. Lee, R. Kuris, D. Koblas, L. Jones. *RFC 1928. SOCKS Protocol Version 5.* March 1996

[Mac00] A. Macaire. *An Open Terminal Infrastructure for Personal Services.* TOOLS Europe 2000, 5-8 June 2000, Le Mont-St-Michel, France

[Roe] Martin Roesch. *Snort - Lightweight Intrusion Detection for Networks* - http://www.snort.org/docs/lisapaper.txt

[SaSa97] R. S. Sandhu, P. Samarati. *Authentication, Access Controls, and Intrusion Detection*, in The Computer Science and Engineering Handbook, pp 1929-1948, 1997.

[SEVA] SEVA project home page - http://www.eurecom.fr/~nsteam/SEVA/

[S-Peer] Texar. *S-Peer.* http://www.s-peer.com/

[SRL] S.X. Sun, S. Reilly, L. Lannom. *Handle System Namespace and Service Definition.* IETF Draft. May 2001.

[SUN] SUN Microsystems. SunScreen Secure Net 3.1, Technical Whitepaper

[TIS] FWTK.ORG *unofficial page on TIS firewall toolkit* - http://www.fwtk.org/main.html

[tpaML] IBM. *Electronic Trading-parter Agreement for E-Commerce. ebXML proposed specification, version 1.06.* http://www.ebxml.org/project_teams/trade_partner/

[W3C00] World Wide Web Consortium. *Extensible Markup Language (XML) 1.0. W3C Recommandation.* http://www.w3.org/TR/2000/REC-xml-20001006

MULTIPLE LAYER ENCRYPTION FOR MULTICAST GROUPS

Alain Pannetrat
Institut Eurecom, Sophia Antipolis, France.
Alain.Pannetrat@eurecom.fr

Refik Molva
Institut Eurecom, Sophia Antipolis, France.
Refik.Molva@eurecom.fr

Abstract   We propose a scalable multicast access control framework targeted for large dynamic groups, where members are added and removed frequently. Access control is provided by the original use of the counter-based block cipher mode of operation to encrypt traffic. This scheme uses intermediary elements in the network that contribute individually to the encryption, providing confidentiality, backward and forward secrecy and also containment, a property that limits the impact of the compromise of member access keys.

Keywords: Multicast Security, Encryption.

1.   INTRODUCTION

This work describes a framework designed to provide access control in a large dynamic multicast[Dee89] group using encryption techniques. Consider, for example, the broadcast of pay-per-view TV over the Internet. In such an application, we need to ensure that only selected recipients are permitted to access the video. The set of recipients may nevertheless be very large and dynamic. As we will see, this complicates the design of an access control protocol. This work does not cover issues such as authentication of multicast data and other multicast security issues. We refer the reader to [HT00] for general presentation of multicast security issues, solutions and challenges.

We focus on a *1-to-n* multicast scenario, where there is one source and many recipients. This approach can be concretely extended to a few sources by using one source as a proxy, however our work does not aim to provide a mechanism where all recipients of the multicast group are potentially also a source. As highlighted in [HC99], the *1-to-n* scenario is well suited to large commercial multicast applications such as pay-per-view broadcasting. Through the remaining of this work we will use the following conventions: we call *recipient*, any entity capable of receiving multicast packets from a certain group, regardless of any cryptographic protection that is applied to the data inside the packets; we call *member*, a recipient who has been given cryptographic keys that enable him to access the content of the received multicast packets.

The primary goal of a multicast access control mechanism is to allow only members to access the content of multicast packets while disallowing other recipients to do so. We assume the existence of one or several entities called membership managers which *add* or *remove* members from the group based on a certain policy (which is beyond the scope of this work). Consequently, the group of members changes *dynamically* through time as members are added or removed from the group. The dynamic nature of the group of members imposes two critical requirements on the access control mechanism: *forward and backward secrecy*. Backward secrecy is defined as the impossibility for a member who is added to the group to access past data while forward secrecy is defined as the impossibility for a member who is removed from the group to access future data. The two main security objectives of a multicast access control framework are thus:

> **R1 - Confidentiality.**
>
> **R2 - Backward and forward secrecy.**

Each time a member is added or removed from the group, the encryption scheme that protects the data needs to be re-keyed to guaranty forward and backward secrecy. The main challenge is to provide a re-keying mechanism for a large multicast group that is *scalable*. As described by the same authors in [MP99], the 3 main scalability requirements of a multicast access control framework can be summarized as:

> **R3 - Processing scalability:** the processing load supported by the entities in the framework should be independent of the group size.
>
> **R4 - Membership scalability:** when a member is added or removed from the group it should not affect the rest of the group.
>
> **R5 - Group-wise scalability:** no operation should require the whole group to be treated as a set of distinct individuals.

The requirements **R1** to **R5** together form the core design goals of a multicast access control framework. However, as highlighted in [MP99], there is an aspect that is often overlooked in multicast access control frameworks: *member compromise*. Generally, the more entities that participate in a security protocol, the greater the chances of a compromise. Thus if we design a multiparty security protocol that scales to a large group of members, we need to be concerned by such an issue: what happens if the access keys of a legitimate member are given to another recipient or stolen by an "hacker", published on a web site or in a newsgroup ? Consequently another requirement should be added to a multicast access control framework:

R6 - Containment: The compromise of one (or several) member(s) should not cause the compromise the entire group.

There are two main directions followed by multicast access control proposals: *key graph* based approaches and *re-encryption tree* based approaches. Key graphs were first proposed by WONG ET AL. [WGL98] and WALLNER ET AL. [WHA98] (see also [CVSP98; MS98; CGI$^+$99]). These authors construct a tree T where each vertex represents a random distinct key and which has as many leaves as members in the group. Each leaf is associated to a member and each member receives a set of keys corresponding exactly to the keys on the path from the root of T to its corresponding leaf. The root of the tree is the key used to encrypt the traffic. When a member is removed from the group the root and all other keys on the path from the root to the leaf representing the departing member are invalidated. These keys are changed and re-broadcasted to the group by encrypting them with other remaining valid keys in the tree in a careful manner such that only the remaining members may update their subset of invalidated keys. We refer the reader to [WGL98] for precise definitions and related algorithms. The main strength of this approach is that it is simple because it does not rely on intermediary elements in the network and it has a reasonable overhead (logarithmic in the size of the group). However, when a member is removed from the group, then all other members are affected since at least one of the keys they hold, the root key, is invalid. As a consequence *all* remaining members in the group need to receive a message to update their access parameters. Thus this scheme does not offer membership scalability (R4). It does not offer containment either, since the compromise of the keys held by a legitimate recipient allows anyone to access the group from anywhere within the scope of the multicast group.

The other approach is based on the same principle that multicast routing protocols use to scale to large groups: involve the intermediary elements in the network. Here, the intermediary elements modify the encryption of the data going through the multicast network, such that the recipient group is partitioned to subsets with different access parameters, thus restricting all scalability issues to an arbitrary small subset of the group. The first proposal to follow this idea was the IOLUS framework

[Mit97] and was used later in IGKMP[HCD00] and in Cipher Sequences [MP99]. We refer to these family of schemes as *re-encryption trees*, since they all use a tree of intermediary elements to perform transformations on the multicast data. Since each subset of the group uses a different key to access the group, there is a dependency between the location of a member in the network and the access parameters that it uses to access the group. If an access key is exposed, it has a limited impact because it is only useful in a subset of the group. Thus, this family of solutions provides containment (R6).

A strong drawback of IOLUS and IGKMP is that they trusts all intermediate elements with the security of the group. A solution to this problem was provided by Cipher Sequences[MP99]. Cipher Sequences allow the intermediate elements to perform security transforms without being trusted, thus satisfying a final requirement for a multicast confidentiality framework:

> **R7 - Limited trust in intermediary elements:** the compromise of some intermediary elements in the network should not compromise the group, or provide access to the protected data.

However, the main drawback of Cipher Sequences is that they rely on asymmetric cryptographic transforms as opposed to other schemes which use classical symmetric encryption. Thus Cipher Sequences cannot be used for bulk data encryption but are instead restricted to key distribution, which typically involves short messages.

1.1   Contribution and outline of this work

The solution we propose here belongs to the family of *re-encryption trees*. The main contribution of this work is the definition of a multicast confidentiality framework that uniquely combines 2 qualities:

1 It satisfies all the requirements described above including R7 like Cipher Sequences.

2 It can be applied for bulk data encryption as in IOLUS or IGKMP by relying on efficient cryptographic techniques.

This work is organized as follows. In the first section, we will look at some interesting cryptographic primitives that we use in our framework. Then, we present our framework based on multiple layers of encryption, or L-layer trees. In the following section we analyze the security of our construction, and discuss its scalability and relate it to the list of seven requirements we presented above. Finally, we present a potential improvement of our scheme with a shorter message expansion.

2. CRYPTOGRAPHIC PRIMITIVES

Our framework uses a multiple key version of the counter based block cipher mode of operation (CTR-Mode) as described in [BDJR97] and can alternatively be constructed with any general stream cipher. CTR-mode has been proposed for standardization to NIST as an official AES mode of operation in [LRW00]. In this section we will briefly recall CTR-mode and present our own multiple key extension that is used in the framework. We denote "$\oplus$" as the binary XOR operation.

2.1 CTRM encryption scheme.

In [BDJR97] BELLARE ET AL. describe and analyze various cipher modes of operation. We will briefly recall their work on the CTR-mode, which we use in our own scheme. Let $f_a(.)$ describe a l-bit pseudorandom permutation such as DES or AES[oST01] where a is the encryption key. The CTR-mode scheme $\text{CTRM}_{f_a} = (\mathcal{K}, \mathcal{E}, \mathcal{D})$ is defined as follows:

- the function $\mathcal{K}$ flips coins and outputs a random k bit key a.

- the function $\mathcal{E}(\sigma, x)$ is defined as:

  split x in n blocks of l bits: $x = x_1, ..., x_n$
  for $i = 1, ..., n$ do $y_i = f_a(\sigma + i) \oplus x_i$.
  return $(\sigma, y_1 y_2 ... y_n)$.
  $\sigma \leftarrow \sigma + n$

- the function $\mathcal{D}(\sigma, y)$ is defined as:

  split y in n blocks of l bits: $y = y_1 y_2 ... y_n$
  for $i = 1, ..., n$ do $x_i = f_a(\sigma + i) \oplus y_i$
  return $x = x_1 x_2 ... x_n$

Note: The state *or counter* σ is maintained by the encryption algorithm across consecutive encryptions with the same key. The decryption algorithm is stateless.

The authors of [BDJR97] have shown that there is a tight reduction between the security of the CTRM-scheme and the security of the primitive block operation f_a (we refer the reader to their work for details).

This scheme has many advantages. First it's paralellizable because the encryption of each block is independent of another. Second, the decryption can under certain circumstances be "prepared" in advance. Since the state is incremented in a predictable way across several messages, it means that the receiver can pre-compute some of the values of f_a to reduce online computations. Finally, this scheme uses the XOR operation which is commutative, a property that we will show to be useful.

2.2   Multiple encryptions.

The commutative nature of the XOR operation makes CTRM$=(\mathcal{K}, \mathcal{E}, \mathcal{D})$ interesting for a special form of multiple encryption. Normally if we encrypt a message x several times with a set of keys $a_1, ..., a_m$ we would compute $(\sigma_m, y) = \mathcal{E}_{a_m}(\sigma_m, ...\mathcal{E}_{a_2}(\sigma_2, \mathcal{E}_{a_1}(\sigma_1, x)...)$ but we proceed slightly differently, by leaving the counters $\sigma_1, ..., \sigma_m$ outside the consecutive encryptions. We define CTRM$^{(m)} = (\mathcal{K}^{(m)}, \mathcal{E}^{(m)}, \mathcal{D}^{(m)})$ with m independent keys as as follows:

- $\mathcal{K}^{(m)}$ chooses m random keys : $a_0, a_1, ..., a_m$.

- $\mathcal{E}^{(m)}_{a_1, a_2, .., a_m}(\sigma_1, ..., \sigma_m, x)$ is defined as:

  split x in n blocks of l bits: $x = x_1, ..., x_n$
  for $i = 1, ..., n$ do $y_i = f_{a_1}(\sigma_1 + i) \oplus ... \oplus f_{a_m}(\sigma_m + i) \oplus x_i$.
  return $(\sigma_1...\sigma_m, y_1 y_2...y_n)$.
  for $j = 1, ..., m$ do $\sigma_j \leftarrow \sigma_j + n$

- $\mathcal{D}^{(m)}_{a_1, ..., a_m}(\sigma_1, ..., \sigma_m, y)$ is defined as:

  split y in n blocks of l bits: $y = y_1 y_2...y_n$
  for $i = 1, ..., n$ do $x_i = f_{a_1}(\sigma_1 + i) \oplus ... \oplus f_{a_m}(\sigma_m + i) \oplus y_i$
  return $x = x_1 x_2...x_n$.

We note immediately that $\mathcal{E}_a = \mathcal{E}_a^{(1)}$ and $\mathcal{D}_a = \mathcal{D}_a^{(1)}$. This form of multiple encryption has the following interesting properties:

Fact 2.1 *For any permutation π of $\{1, ..., m\}$ we have*

$$\mathcal{E}_{a_{\pi(1)}, ..., a_{\pi(m)}}(\sigma_{\pi(1)}, ..., \sigma_{\pi(1)}, x) = \mathcal{E}_{a_1, a_2, .., a_m}(\sigma_1, ..., \sigma_m, x)$$

and

$$\mathcal{D}_{a_{\pi(1)}, ..., a_{\pi(m)}}(\sigma_{\pi(1)}, ..., \sigma_{\pi(1)}, y) = \mathcal{D}_{a_1, a_2, .., a_m}(\sigma_1, ..., \sigma_m, y)$$

This is a natural consequence of the commutativity of the XOR binary operation.

Fact 2.2 *Given a message x if we compute*
$\{\sigma_1, ..., \sigma_{(m-1)}, y\} \leftarrow \mathcal{E}^{(m-1)}_{a_1, ..., a_{(m-1)}}(\sigma_1, ..., \sigma_{(m-1)}, x)$ *and* $\{\sigma, z\} \leftarrow \mathcal{E}_a^{(1)}(\sigma, y)$
then we have $\{\sigma_1, ..., \sigma_{(m-1)}, \sigma, z\} = \mathcal{E}^{(m)}_{a_1, ..., a_{(m-1)}, a}(\sigma_1, ..., \sigma_{(m-1)}, \sigma, y)$. *A similar result holds for $\mathcal{D}^{(m-1)}$ and $\mathcal{D}^{(1)}$.*

We also recall a classical property of multiple commutative encryptions that applies to our scheme [MvOV96, Chapter 7]:

Fact 2.3 *When a message is encrypted with m keys as described above it is at least as secure as anyone of the individual encryptions.*

3.   L-LAYER ENCRYPTION TREES

3.1   Definitions

To describe our *1-to-n* multicast encryption framework, we view the multicast network as a tree with the following elements:

root: The root represents the source generating data to be securely distributed to members of the group.

intermediary: An intermediary describes any node in the tree besides the root and the leaves. An intermediary element is either a multicast enabled router or proxy with embedded encryption capabilities.

leaf: The leaf represents a member, or a cluster of members receiving data from the same intermediary. These elements are expected to be physically close to each other, for example on the same LAN with a common IGMP[Fen97] router.

Our approach to secure multicast can be summarized as follows. The root produces data, encrypts it and forwards it to the multicast network. intermediaries receive data, modify the encryption and forward the result to other intermediaries until it reaches a leaf. Finally, the members in the leaves decrypt the data transmitted by the source. Since each intermediary element changes the encryption of the data, each leaf will require different access parameters to access the content, which provides a form containment as described in the introduction.

3.2   Construction

We call a tree T a *singular leaf* tree if each leaf in T has a distinct unique parent. We define a function $Depth(N)$ which for a node N returns its depth in the tree, where $Depth(root) = 0$, and we define the function $Parent(N, L)$ which returns the Lth parent of node N if it exists or $\emptyset$ otherwise.

We call *"L-layer tree"* the association of a multicast singular leaf tree network with a set of cryptographic transformations designed to protect the distributed data with a varying set of L layers of encryption. We associate a set of keys to the tree to perform CTRM encryptions as described previously, taking advantage of the commutative nature of the encryption scheme. Let T be a tree without sibling leaves with n intermediaries. We associate a set of $n + L$ different encryption keys $[K_1, ..., K_{L+n}]$ to the tree as follows:

root: The root receives the encryption keys $[K_1, ..., K_L]$.

intermediaries: The n intermediaries receive each a distinct key from the set $[K_{L+1}, ..., K_{L+n}]$. For example if we number the intermediary arbitrarily from 1 to n we can associate key K_{L+i} to intermediary number i. We call this key the intermediary's *encryption* key. Each intermediary N receives a secondary key K', which we will call *decryption* key, as follows:

if $Depth(N) < L$ then $K' \leftarrow K_{Depth(N)}$.

else $K' \leftarrow$ (the *encryption* key of $Parent(N, L)$)

leaves: The leaves each receive L keys. To clarify the notation we will call these keys $X_1, ..., X_L$, A leaf N receives these keys as follows:

for $i = 1, ..., L$ do

(1) if $Parent(N, i) = \emptyset$ then $X_{(L-i+1)} \leftarrow K_{Depth(N)+i-1}$

(2) else $X_{(L-i+1)} \leftarrow$ (the encryption key of $Parent(N, L)$)

Line (1) shows that if the leaf does not have an ith parent then it gets one of the encryption keys used by the root and line (2) shows that if it does have an ith parent then it gets the encryption key of that parent.

This key assignment may seem somewhat complex but in fact it's governed by two simple principles:

- Each intermediary gets its own encryption key and the encryption key of its Lth parent.

- Each leaf gets all the encryption key of its parents of level L down to 1.

The complexity only appears in the algorithm for nodes or leaves that are not deep enough in the tree to fully apply the previous two rules. In such a case, the otherwise missing keys are taken from the root. If we focus our attention on a single path of the tree extending from the root to a leaf, we can see that each key used on a node in a path is used once as an encryption key and once as a decryption key, as illustrated on figure 1.

An example of our key assignment algorithm is shown on figure 2 for a 4 layer tree.

3.3   Data Distribution

Once the tree is constructed, its components operate as follows:

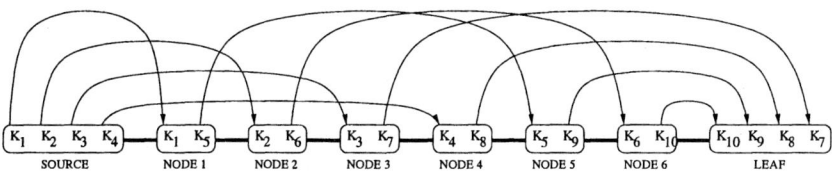

Figure 1. Key distribution on a single path in a 4 layer tree.

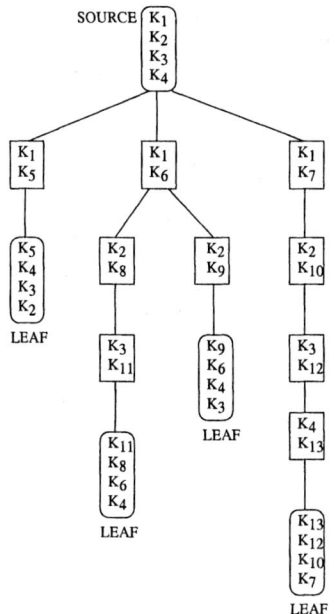

Figure 2. Key distribution on a 4 layer tree.

root: The root or source encrypts a message M by computing $(\sigma_1, ..., \sigma_L, C) = \mathcal{E}^{(L)}_{K_1, ..., K_L}(\sigma_1, ..., \sigma_L, M)$ and sends the result to its children nodes in the tree.

intermediaries: Each intermediary N receives an encrypted message $(\sigma_1, ..., \sigma_L, C)$. The intermediary N performs the following operations:

1 Suppress a layer of encryption: $C' \leftarrow \mathcal{D}^{(1)}(\sigma_1, C)$.

2 Add a new of encryption: $(\tau, C'') \leftarrow \mathcal{E}^{(1)}(\tau, C')$ where τ is the internal counter of N.

3 Let $\tau_L \leftarrow \tau$ and $\tau_i \leftarrow \sigma_{(i+1)}$ for $i = 1, ..., (L-1)$. Send $(\tau_1, \tau_2, ..., \tau_L, C'')$ to the children nodes.

leaves: The leaves receive an encrypted message $(\sigma_1, \sigma_2, ..., \sigma_L, C)$ that they decrypt by computing $M = \mathcal{D}^{(L)}_{X_1, ..., X_L}(\sigma_1, ..., \sigma_L, C)$.

If we recall the construction of our tree, we see that the keys are distributed to make the above algorithm work: each key used to encrypt the data is used later as a decryption key. The source and the leaves both perform L-encryptions and L-decryption. Intermediaries use Property 2 to first transform an L-encryption to a $(L-1)$ encryption, then using the same property, they transform the $(L-1)$-encryption back into an L-encryption. The combination of Property 2.1 and 2.2 allows us to decrypt a layer regardless of the order in which the encryptions were done.

As an example, we will examine how our data distribution algorithm is applied on the path of the 4-layer tree of figure 1, where C^r denotes the encryption of M at stage r in the algorithm:

Source: Computes and sends
$$(\sigma_1, \sigma_2, \sigma_3, \sigma_4, C^0) \leftarrow \mathcal{E}^{(m)}_{K_1, K_2, K_3, K_4}(\sigma_1, \sigma_2, \sigma_3, \sigma_4, M)$$

Node 1: Suppresses a layer $(\sigma_2, \sigma_3, \sigma_4, C^1) \leftarrow \mathcal{D}^{(1)}_{K_1}(\sigma_1, C^0)$.
   Then it computes and sends $(\sigma_2, \sigma_3, \sigma_4, \sigma_5, C^2) \leftarrow \mathcal{E}^{(1)}_{K_5}(\sigma_5, C^1)$.

Node 2: Suppresses a layer $(\sigma_3, \sigma_4, \sigma_5, C^3) \leftarrow \mathcal{D}^{(1)}_{K_2}(\sigma_2, C^2)$.
   Then it computes and sends $(\sigma_3, \sigma_4, \sigma_5, \sigma_6, C^4) \leftarrow \mathcal{E}^{(1)}_{K_6}(\sigma_6, C^3)$.

Node 3: Suppresses a layer $(\sigma_4, \sigma_5, \sigma_6, C^5) \leftarrow \mathcal{D}^{(1)}_{K_3}(\sigma_3, C^4)$.
   Then it computes and sends $(\sigma_4, \sigma_5, \sigma_6, \sigma_7, C^6) \leftarrow \mathcal{E}^{(1)}_{K_7}(\sigma_7, C^5)$.

Node 4: Suppresses a layer $(\sigma_5, \sigma_6, \sigma_7, C^7) \leftarrow \mathcal{D}_{K_4}^{(1)}(\sigma_4, C^6)$.
Then it computes and sends $(\sigma_5, \sigma_6, \sigma_7, \sigma_8, C^8) \leftarrow \mathcal{E}_{K_8}^{(1)}(\sigma_8, C^7)$.

Node 5: Suppresses a layer $(\sigma_6, \sigma_7, \sigma_8, C^9) \leftarrow \mathcal{D}_{K_5}^{(1)}(\sigma_5, C^8)$.
Then it computes and sends $(\sigma_2, \sigma_3, \sigma_4, \sigma_5, C^{10}) \leftarrow \mathcal{E}_{K_9}^{(1)}(\sigma_9, C^9)$.

Node 6: Suppresses a layer $(\sigma_7, \sigma_8, \sigma_9, C^{11}) \leftarrow \mathcal{D}_{K_6}^{(1)}(\sigma_6, C^{10})$.
Then it computes and sends $(\sigma_7, \sigma_8, \sigma_9, \sigma_{10}, C^{12}) \leftarrow \mathcal{E}_{K_{10}}^{(1)}(\sigma_{10}, C^{11})$.

Leaf: Decrypts the message $M \leftarrow \mathcal{D}_{K_7, K_8, K_9, K_{10}}^{(m)}(\sigma_7, \sigma_8, \sigma_9, \sigma_{10}, C^{12})$.

3.4 Membership Management

After describing how members access the multicast content in the previous section, we will now turn our attention to the addition and removal of members in our framework.

Add: When a recipient M wants to be added to the group, he contacts a membership manager (MM) with an authenticated secure channel. If M is allowed to access the group, then there are 2 possible scenarios:

1 M is already physically in an existing leaf F: the MM sends an authenticated secure message to the parent intermediary P of F, to change the encryption key K of P to a new value K'. Then the new key K' is sent to all the members of the same leaf and to the new member M.

2 M is not in an existing leaf: the tree is expanded to create a new leaf for M. The corresponding keys are distributed to the new intermediaries and M.

Remove: When a member needs to be removed from the group, the MM sends an authenticated secure message to the parent intermediary P of F, to change the encryption key K of P to a new value K'. Then the new key K' is sent to all the members of the same leaf except M. If the leaf is empty because the last member left, than after a certain delay, we may remove unused intermediaries from the tree.

The leaf holds L keys and needs all of them to access the data. Thus, changing just one of them when we add or remove a member provides us with forward and backward secrecy (R2).

3.5 Distributed Membership Management.

This scheme also lends itself to a certain form of decentralized key management. The tree of intermediary elements can be managed by a hierarchy of membership managers. It follows from our construction in section that an individual membership manager only needs to know L extra keys to manage a subtree on its own. More precisely, if a membership manager is selected to manage a subtree consisting of an intermediary N and all its descendants in the tree, then it needs the to know the set $Y_1, ..., Y_L$ of keys defined as follows:

    for $i = 1, ..., L$ **do**

        if $Parent(N, i) = \emptyset$ **then** $Y_i \leftarrow K_{Depth(N)+i-1}$

        else $Y_i \leftarrow$ (the encryption key of $Parent(N, i)$)

In turn a membership manager may delegate the management of some of its own subtrees to several other membership managers.

4. SECURITY REQUIREMENTS

4.1 Encryption

To discuss the security of our construction, we will first look at *one-layer* trees before we study the general case. One layer trees are conceptually very simple, since they only use the original CTRM encryption algorithm. The source has a key K_1 and uses it to encrypt data to be sent to its children. The intermediaries decrypt the data with the key K_j that their parents used to encrypt the data and use their own key K_i to encrypt the data again for their children. The leaves use a single key to access the data. A one-layer tree is quite similar to the IOLUS framework [Mit97], and it shares one of the drawbacks of that framework: each intermediary is trusted to access the cleartext data. For now however, let's examine the security off a one-layer tree while making the hypothesis that the intermediary elements are secure.

The individual links are secured by the CTRM encryption algorithm. In our framework, an adversary has the ability to observe several links and thus the same message encrypted under different keys. We can even imagine that the adversary may modify or input new messages at different points in the tree to try to break the security of the system. In a recent work evaluating the security of public key cryptosystems in the multiuser setting [BBM00], BELLARE ET AL. have shown essentially that if a public key cryptosystem is secure in the sense of indistinguishability, then it implies the security of the cryptosystem in the multiuser setting, where related messages are encrypted under different keys. We refer the reader to their work for further details [BBM00]. Though their work was targeted at public key cryptosystems, their results can be applied to the private key setting, and since the CTRM encryption is secure in the sense of "indistinguishability" under chosen plaintext attacks[BDJR97], we can assert the security of the whole tree by using the results of [BBM00].

Now for L-layer trees, property 2.3 tells us that they are at least as secure as a 1-layer tree if no intermediary is compromised. But, the advantage of a L-layer tree is that it remains secure even if some nodes are compromised, more precisely:

Proposition 4.1 Let $B = B_1, ..., B_p$ define a set of p compromised intermediaries in a L-layer tree. The tree remains secure as long as there exists a constant $c \in \{0, ..., L-1\}$ such that $Depth(B_i) \neq c \bmod L$ for all $i \in \{1, ..., p\}$.

Proof. This property derives from the arrangement of the keys in the tree. Let $B = B_1, ..., B_p$ define a set of compromised intermediaries in a L-layered tree T such that there exists a constant $c \in \{0, ..., L-1\}$ verifying $Depth(B_i) \bmod L \neq c$ for all $i \in \{1, ..., p\}$. From the tree T we can extract a subtree $\overline{T}$ iteratively, as follows:

Notations:
Let N_0 define the root of T.
Let $\overline{N}_0$ define the root of $\overline{T}$, and let $\{\overline{N}_1, ..., \overline{N}_q\}$ define the intermediary nodes of $\overline{T}$.

Construction:
$\overline{N}_0 \leftarrow N_0$
Select $\{N_1, ..., N_q\}$, the set of intermediaries N_i of T which verify $Depth(N_i) = c \bmod L$.
$\{\overline{N}_1, ..., \overline{N}_q\} \leftarrow \{N_1, ..., N_q\}$
for $i = 1, ..., q$ **do**
  if $Depth(N_i) = c$ **then**
    connect $\overline{N}_i$ to $\overline{N}_0$ in $\overline{T}$.
    let $\overline{Z}$ be the concatenation of all leaves $Z_k \in T$ such that $Depth(Z_k) \leq c$.
    if $\overline{Z} \neq \emptyset$ **then** connect $\overline{Z}$ to $\overline{N}_0$ in $\overline{T}$.
  else
    let $N_j = Parent(N_i, L)$.
    connect $\overline{N}_j$ to $\overline{N}_i$ in $\overline{T}$.
    let $\overline{Z}$ be the concatenation of all leaves $Z_k \in T$ such that
      $(Parent(Z_k, r) = N_i$ and $j \leq L)$.
    if $\overline{Z} \neq \emptyset$ **then** connect $\overline{Z}$ to $\overline{N}_i$ in $\overline{T}$.

The tree $\overline{T}$ represents a 1-layer tree such that none of its intermediaries $\{\overline{N}_1, ..., \overline{N}_q\}$ hold a key in common with any of those distributed to the compromised set B. Thus since there exists an independent 1-layer tree between the root and the leaves, the encryption of data in the tree remains secure (R1).   ◇

Corollary 4.2 An obvious implication of this property is that an L-layer tree can at least withstand the compromise of any set of less than L intermediaries.

4.2   Containment

In singular parent trees, no leaf shares its direct parent with another leaf, thus each leaf receives data that is encrypted with at least one layer of encryption that is distinct from any other leaf. This distinct layer of encryption is generated by the parent intermediary node of the leaf. Thus if $\mathcal{L}$ is a leaf, then no collusion of any group of leaves $\{\mathcal{L}_1, ..., \mathcal{L}_p | \mathcal{L}_i \neq \mathcal{L}, i \in \{1, ..., p\}\}$ can break the encryption of data received in the leaf $\mathcal{L}$. An adversary in a leaf $\mathcal{L}$ who compromises the keys in a set of leaves $\{\mathcal{L}_1, ..., \mathcal{L}_p | \mathcal{L}_i \neq \mathcal{L}, i \in \{1, ..., p\}\}$ cannot use this information to access the data in his own leaf. Thus having a *singular leaf* tree is a sufficient condition to ensure a secure dependence between the location of a recipient in the network and the keys used to decrypt the received multicast data. This secure dependency provide containment (R6) since the exposure of a key will only be useful to an adversary within the same leaf and will not affect the security of the whole group.

There is no containment within a leaf, all the recipients that are physically in the same leaf use the same key to access the data, thus exposure of keying material in one leaf allows other members of the same leaf to access the data. However, unless there is a form of hardware access control installed directly on each recipient, providing containment in within a leaf is very hard: ultimately, it's difficult to stop or even detect if a member rebroadcasts decrypted data to other local recipients that are not members themselves.

4.3   Hybrid attacks

The current framework may face more complex attacks which are a combination of both the compromise of leaves and intermediary elements:

Membership extensions:.   If a member M in a leaf $\mathcal{L}$ takes full control of the direct parent P of $\mathcal{L}$, it can monitor key changes in P. If the membership manager decides to remove M from the group, it will change the key K held by P and send the new updated key K'

to the other remaining members in $\mathcal{L}$ as well as P. As a consequence the removed member M will still be able to stay in the group because it learn the new value K' from P. This means that we lose forward secrecy. Recovering from such a compromise requires a key change in the parent P' of the compromised node P, which in turn requires all the leaves that have P' as an ancestor to be updated.

Containment failures:. Assume that two leafs $\mathcal{L}_1$ and $\mathcal{L}_2$ of same depth in the tree share a common ancestor node P in the tree which verifies $|Depth(\mathcal{L}_1) - Depth(P)| \leq L$. In that situation, the members in $\mathcal{L}_1$ and the members in $\mathcal{L}_2$ have $k < L$ decryption keys in common. If a member M_1 in $\mathcal{L}_1$ compromises the $L - k$ first parents of $\mathcal{L}_2$ than M_1 will know enough information to generate the set of L keys used in $\mathcal{L}_2$, by combining the k common keys with the $L - k$ compromised keys. This attacks breaks the containment property of the scheme for two leafs that are at the same depth in the tree.

5. SCALABILITY REQUIREMENTS

The processing load supported by each entity in the tree is not proportional to the group size. For the leaves and the root it depends on the parameter L which defines the number of layers in the tree, while intermediaries always perform a single decryption and a single encryption regardless of the number of layers. Thus this framework offers processing scalability (R3).

When a member is added or removed from the group, the key update remains local and only affects a leaf at a time. The number of elements in a leaf is not a scalability factor itself, because we can simply create more branches in the tree to cope with leaves that get too large. Consequently, our framework provide membership (R4) and group-wise scalability (R5).

One of the main differences in terms of scalability between key graph [WGL98] approaches and intermediary based approaches like ours, is membership scalability. In key graph approaches, the departure of a member requires the whole group to receive a message to update its access keys.

6. REDUCING EXPANSION

In the $CTRM^{(m)}$ scheme, the encryption of a message results in an expansion of $m.|\sigma_i|$ bytes where $|\sigma_i|$ represents the size in bytes of the state value. We could use a single state chosen by the source and common to all layers of encryption as well as all elements in the tree. In other words we would rewrite the encryption algorithm as follows:

$\mathcal{E}^{*(m)}_{a_1,a_2,...,a_m}(\sigma, x)$:

split x in n blocks of l bits: $x = x_1...x_n$

for $i = 1, ..., n$ **do** $y_i = f_{a_1}(\sigma + i) \oplus ... \oplus f_{a_m}(\sigma + i) \oplus x_i$.

return $(\sigma, y_1 y_2 ... y_n)$

$\sigma \leftarrow \sigma + n$

The intermediaries would use the same σ to both encryption and decryption operations. The algorithm would be simplified and the ciphertext size would be independent of the number of layers in the tree. In such a case, however, proving the security of the scheme is an open problem since the intermediaries are now stateless and cannot be modeled as independent encrypting devices, which was a requirement of the security proof found in [BBM00] upon which we relied for our scheme.

7.   CONCLUSION

Using intermediary elements in the network we have constructed a scalable framework for multicast access control. This framework offers interesting properties such as containment, and limited trust in the intermediary elements of the network. It shows some vulnerabilities when both members and intermediary elements in the network are compromised, in particular the direct parent intermediary of a leaf in the tree.

Interesting applications of this scheme are not necessarily limited to IP-Multicast. Consider for example the use of this scheme for content distribution in next generation mobile networks. Our scheme would allow a provider to send protected data to its clients even if they are roaming in foreign or "less trusted" networks. Moreover, the risk of key piracy found for example in European Digital Video Broadcasting would be limited by the containment property of our scheme.

References

M. Bellare, A. Boldyreva, and Silvio Micali. Public-key encryption in a multiuser setting: Security proofs and improvements. In *Eurocrypt 2000*, volume LNCS 1807, pages 259–274. Springer Verlag, 2000.

Mihir Bellare, Anand Desai, E. Jokipii, and Phillip Rogaway. A concrete security treatment of symmetric encryption. In *IEEE Symposium on Foundations of Computer Science*, pages 394–403, 1997.

R. Canetti, J. Garay, G. Itkis, D. Micciancio, M. Naor, and B. Pinkas. Multicast security: A taxonomy and some efficient constructions. In *Proceedings of IEEE Infocom'99*, 1999.

G. Caronni, M. Valdvogel, D. Sun, and B. Plattner. Efficient security for large and dynamic multicast groups. In *Proceedings of IEEE WETICE'98*, 1998.

Steve E. Deering. RFC 1112: Host extensions for IP multicasting, Aug 1989.

W. Fenner. Internet group management protocol, version 2. Request For Comments 2236, November 1997. see also draft-ietf-idmr-igmpv3-and-routing-01.txt for IGMP v3.

H. Holbrook and D. Cheriton. IP multicast channels: EXPRESS support for large-scale single-source applications. In *Proceedings of ACM SIGCOMM'99*, Harvard University, September 1999. ACM SIGCOMM.

Thomas Hardjono, Brad Cain, and Naganand Doraswamy. Intra-domain group key management protocol. Internet-Draft, work in progress, February 2000.

T. Hardjono and G. Tsudik. IP multicast security: Issues and directions. *Annales des Telecommunications*, to appear in 2000.

H. Lipmaa, P. Rogaway, and D. Wagner. Comments to NIST concerning AES modes of operation: CTR-Mode encryption. In *NIST First Modes of Operation Workshop*, Baltimore, Maryland, USA, October 20 2000.

Suivo Mittra. Iolus: A framework for scalable secure multicasting. In *Proceedings of the ACM SIGCOMM'97 (September 14-18, 1997, Cannes, France)*, 1997.

Refik Molva and Alain Pannetrat. Scalable multicast security in dynamic groups. In *Proceedings of the 6th ACM conference on Computer and Communications Security*, pages 101–112, Singapore, November 1999. Association for Computing Machinery.

David A. McGrew and Alan T. Sherman. Key establishment in large dynamic groups using one-way function trees. Technical report, TIS Labs at Network Associates, Inc., Glenwood, MD, 1998.

Alfred Menezes, Paul C. van Oorschot, and Scott A. Vanstone. *Handbook of Applied Cryptography*. CRC Press, 1996.

National Institute of Standards and Technology. Advanced Encryption Standard, 2001.

C. K. Wong, M. Gouda, and S. S. Lam. Secure group communications using key graphs. In *ACM SIGCOMM 1998*, pages 68–79, 1998.

Debby M. Wallner, Eric J. Harder, and Ryan C. Agee. Key management for multicast: Issues and architectures. Internet draft, Network working group, september 1998, 1998.

ACCESS CONTROL, REVERSE ACCESS CONTROL AND REPLICATION CONTROL IN A WORLD WIDE DISTRIBUTED SYSTEM

Bogdan C. Popescu
Vrije Universiteit, Amsterdam
bpopescu@cs.vu.nl

Lt. Col. Dr. Chandana Gamage
Sri Lanka Army Headquarters, Colombo, Sri Lanka *
chandag@cse.mrt.ac.lk

Andrew S. Tanenbaum
Vrije Universiteit, Amsterdam
ast@cs.vu.nl

Abstract   In this paper we examine several access control problems that occur in an object-based distributed system that permits objects to be replicated on multiple machines. First, there is the classical access control problem, which relates to which users can execute which methods. Second, we identified a reverse access control problem, which concerns which replicas can execute which methods for authorized users. Finally, there is the issue of how updates are propagated securely from replica to replica. Our solution uses roles and preserves the scalability needed in a world-wide distributed system.

Keywords: Distributed Systems, Replicated Objects, Security, Access Control, Digital Certificates.

---

* Work completed while at Vrije Universiteit

1. SECURITY IN DISTRIBUTED SYSTEMS

Security in distributed systems differs from operating system security by the fact that there is no central, trusted authority that mediates interaction between users and processes. Instead, a distributed system usually runs on top of a large number of loosely coupled autonomous hosts. Those hosts may run different operating systems, and may have different security policies, which can be enforced in different ways by careless, or even malicious administrators.

The popular trend in distributed systems is to encapsulate functionality as objects and provide mechanisms for their location, migration, persistence, as well as for remote method invocation. CORBA [1] [2], DCOM [5], and Legion [6] are examples of distributed systems using this paradigm. Each of them handles security in its own way, and the main objectives are authenticating the communicating parties, protecting network traffic, enforcing access control policies on the object's member functions, delegating rights and respecting site-specific security concerns. There is one feature these systems have in common: all of them support only nonreplicated objects. This makes it easy to implement access control mechanisms for individual objects, since such mechanisms would have to be enforced in only one point, namely at the host where the object resides.

Globe [14], is a wide-area distributed system based on *distributed shared objects* (DSO). The notion of a DSO stresses the property that objects in Globe are not only shared by multiple users, but also physically replicated at possibly thousands of hosts [4].

This paper addresses the security issues arising from the physical replication of objects. In particular, we focus on how access control policies can be implemented in a system where those policies need to be enforced at a number (possibly very large) of distinct locations, with various degrees of trustworthiness.

The security challenges posed by Globe come from the fact that millions of users can invoke methods on any of the possibly thousands of replicas of a highly distributed object. It makes sense to assume that in such an object, there is a trust hierarchy (some of the replicas run on hosts that are more trusted than others), and this translates into what kind of actions the various replicas should be allowed to execute. In nonreplicated systems, the **access control problem** is how to prevent unauthorized users from invoking methods for which they have no rights. With replicated objects, we also need to prevent legitimate users from sending security critical requests to replicas that are not trusted enough to execute them (we call this the **reverse access control problem**). Finally, the **replication control problem** deals with enforcing a security policy on the way replicas exchange state update messages in order to keep the DSO's state consistent.

The rest of the paper is organized as follows: Section 2 gives an overview of the Globe system, the internal structure of a DSO, and how replicas interact to implement the DSO functionality. The following three sections deal with the access control problem, reverse access control problem and replication control problem respectively. Section 6 gives a quick outline on how one would implement the solutions we proposed for these problems, and Section 7 gives an application example. Finally, Section 8 concludes and outlines areas of future research.

2. THE GLOBE SYSTEM

A central construct in the Globe architecture is the distributed shared object (DSO). A DSO is built from a number of **local objects** that reside in a single address space and communicate with local objects in other address spaces.

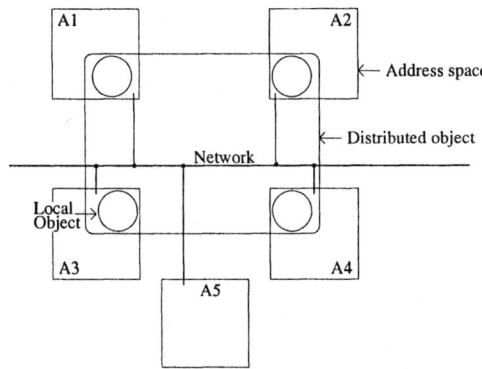

Figure 1. A Globe DSO replicated across four address spaces

Some of the local objects (possibly all of them, depending on the replication strategy) can store all or part of the DSO's state. A local object that stores some part of the DSO's state is called a **replica**.

All the replicas that are part of a DSO work together to implement the functionality of that DSO. Replicas consist of the code for the application, the state they store, and the distribution mechanism. The internal structure of a local object is shown in Figure (2). The distribution mechanism enables transparent distribution and replication of objects, hiding these details from application developers.

The semantics subobject contains the code that implements the functionality of the DSO. This is the only subobject that needs to be written by the application developer.

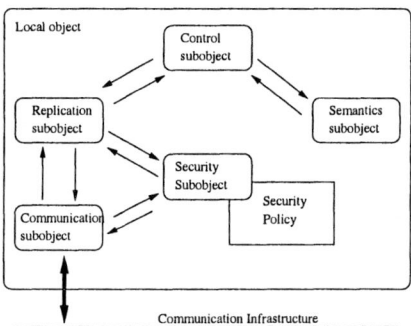

Figure 2. The internal structure of a local object. The arrows indicate the possible interactions between the subobjects

The communication subobject is responsible for the communication between local objects residing in different address spaces. It hides the network communication aspects from all the other subobjects.

The replication subobject is responsible for keeping the state of the local object consistent according to the per-DSO replication strategy. This is accomplished by exchanging state update messages with the replication subobjects of the other replicas that make up the DSO. The replication subobject is also responsible with providing the user with the view of a logical non-replicated object. This is accomplished by transforming method invocations that cannot be handled locally into remote requests, and sending those requests to the replicas that can handle them.

The control subobject is in general system-provided. Its job is to take care of invocations from client processes on the host where the local representative resides and to mediate the interaction between the semantics subobject and the replication subobject.

The security subobject [8] is responsible for enforcing the object's security policy at the local representative level. This is done by having the security subobject mediating the communication flow between the other local subobjects. Prior to passing a protocol message to a lower or upper level in the subobject hierarchy, a subobject passes it to the security subobject. The security subobject can apply security measures to protocol messages (encrypting data before sending it to the network, for example) or can even prohibit the continuation of the protocol if that would be against the DSO's security policy (for example when an un-authorized user tries a remote method invocation on a replica).

In the secure version of Globe, all actions (creating a DSO, running a DSO replica, invoking a method ...) are done by **principals**. A principal can be any entity (human user, group of users, institution ...) that has a public key and a digital certificate certifying that key. We assume the existence of a Public Key Infrastructure used for the distribution and management of these certificates.

Each Globe DSO must have an **owner**, who is the principal that is in charge of the administration and security policy for that object. The **object's public key** is the public key of the object's owner.

Any DSO replica must be run by some principal. The **replica's public key** is the public key of the principal that runs that replica. The replica's principal may or may not be the same as the object owner.

For a DSO, any principal that calls the object's methods is a **user** for that object. Such a user must have a local representative of the object installed in its address space.

It is important to understand that users and replicas are orthogonal concepts in our architecture. Replicas are the building blocks of the DSO. They store parts of the DSO's state and interact according to a replication protocol in order to implement the DSO's functionality. Users are external world entities that invoke the DSO's methods. They see the DSO as a single logical object through the DSO's local representative installed in their address space (this representative could be a replica, if any part of the DSO's state is stored in it, but that is dependent on the replication policy).

3.   THE ACCESS CONTROL PROBLEM

As stated before, for a given application, the access control problem is how to restrict an user to execute only those operations allowed under the application's security policy with respect to that user. For Globe, access control policies are defined on a per-DSO basis, so each object can have its own security policy.

The access control problem has been solved in different ways for the various existing distributed systems. One approach is to associate resources with Access Control Lists (ACLs) [10]. Such a list would simply enumerate all the individuals allowed to use that resource. If that resource is an object, this can be fine-grained by specifying which methods are accessible to each individual. Another possible approach is to use capabilities. A capability is a protected identifier that both identifies an object and specifies access rights to be allowed to the accessor that possesses the capability [3].

ACLs are not suited for implementing access control in systems with world-wide replication of objects. For a heavily replicated DSO, the ACL would have to be stored by each of its thousands of replicas. Even in a distributed system where objects are not replicated, the size of an ACL is proportional to the number of users. In a system like Globe, the storage needed for the ACL would also be proportional to the number of replicas for that object. And what makes the situation even worse is the need to have strong consistency among all the ACLs stored by the various replicas worldwide.

Capabilities are also unsuited for implementing access control in a system like Globe because they suffer from the so-called confinement problem: it is hard to prevent a capability from being passed from one user to another without the object's approval. This problem can be solved in centralized systems by not allowing users to directly manipulate their capabilities. Instead, they are stored by the (trusted) underlying system and presented to objects on users' behalf. Since there is no underlying trusted system in Globe, this technique cannot be applied.

Role Based Access Control (RBAC) [11] represents another approach to the access control problem, and has been the focus of intense research in the past few years [12], [9]. The main idea behind RBAC is that permissions are associated with roles instead of users. Roles are abstractions that group entities with equivalent security properties for an application.

In Globe, we identify a role as a subset of all the methods offered by a DSO. If there are N methods, in theory there are $2^N - 1$ potential roles, although most of these are not likely to be useful. This two-level scheme (users to roles and roles to permissions) greatly simplifies the management of access control lists. First of all, we expect many fewer roles than users (since roles group users with the same rights). This will result in much more compact access control lists. Second, remember that one major problem with ACLs in a worldwide system with replicated objects is keeping them consistent, since users can be granted or revoked rights at any time. With RBAC, this is done by assigning that user in a new role without modifying the ACL. In this way, ACLs can remain largely static while user roles are dynamically managed.

A legitimate question to raise is how does a DSO keep track of which roles have been assigned to which users? If this is kept as a list, we encounter the same problems as for ACL's, namely how to make such a list available to all replicas, while keeping it consistent. The solution is to use "role certificates". These certificates bind the user's public key with the role it has been assigned, and are signed by the object's owner.

When designing a DSO security policy, the object's owner first has to identify all the meaningful roles for that object. This is accomplished by careful examination of the application being implemented; some roles are needed to represent the various types of clients (for example a banking application may need to differentiate between regular members, gold card members and platinum card members). Other roles are needed to model the administrative hierarchy for the application - using the same banking example - we may need teller roles that are allowed to process withdrawals/deposits and manager roles who are allowed to approve loans. At the end, all the selected roles form the **user role set** for that object.

Now, we can implement the access control scheme for a given DSO using the user role set we identified for that DSO: with each user role we associate a bit vector encoding the methods that role is allowed to invoke (for example if a user role is allowed to invoke methods M_2 and M_5, its vector will be 01001...). Grouping the vectors for all the user roles for a DSO, we obtain the **access control matrix** for that DSO, which is stored in the security subobject. In this matrix, the first row contains the bit vector for the first role, the second row contains the bit vector for the second role, and so on. The DSO's replicas get this matrix from the object's owner when they are created. When a user invokes a method on a given replica, the security subobject on that replica first checks the access control matrix to make sure the user's role is allowed to call that method.

4. THE REVERSE ACCESS CONTROL PROBLEM

The reverse access control problem is how to prevent legitimate users from sending service requests to applications not entitled to provide those services. Those malicious servers can fool legitimate users into believing that they have performed some action on behalf of them, when in fact they are not even allowed to perform it under the application's security policy

For distributed systems where objects are not physically replicated, the reverse access control problem is reduced to whether or not one trusts an object to perform a given action. This can be mediated by security policies based on the location of the object, or on the entity that owns it, and so on. The situation dramatically changes with worldwide replication: an object can be replicated on thousands of different systems. It makes sense to assume that only few of these replicas should be trusted to perform the most security sensitive operations provided by the object (like changing the object's state). The rest of the replicas are probably there for performance reasons, acting as caches for example.

What we need is a systematic and scalable way of describing which replicas are allowed to execute which methods for a given DSO. This can be tackled by examining that DSO and answering the question "why does this object need to be distributed?" One answer is that the functionality the object implements is suited to be divided among many parts – the replicas – spread across the network. When running, each of these replicas will implement part of what the DSO is supposed to do; we could divide the set of all replicas of a DSO into disjoint subsets with equivalent functionality; such replicas will have the same role in implementing what the DSO is supposed to do. We can see that the disjoint subsets we described, closely follow the replication strategy used for the object. For example, a master-slave replication strategy would create two such sets - the masters and the slaves. We name such a disjoint subset a **replication role**. All the replication roles identified for a given DSO form the **replication role set** for that DSO.

We now claim that if one replica in a given replication role is allowed to execute a client request for a method M of the object, then all the other replicas in that role should also be allowed to execute that request. Remember that a replication role is the set of all replicas with equivalent functionality. If some replicas in the role are allowed to execute method M, while others are not, then those replicas would differ in functionality (some are allowed to execute M and some not), and this contradicts the assumption they are in the same role.

Now, we can implement the reverse access control scheme for a given DSO using the replication role set we identified for that DSO: with each replication role we associate a bit vector encoding the client method requests that role is allowed to execute (for example if a replication role is allowed to serve methods M_2 and M_5, its vector will be 01001...). Grouping the vectors for all the replication roles for a DSO, we obtain the **reverse access control matrix** for that DSO. The DSO's users get this matrix from the object's owner when they set up their local representative, and store it in the security subobject of that representative. When a user wants to invoke one of the object's methods, its security subobject will have to search in the matrix to find a role allowed to execute it.

Finally, a replica needs a **role certificate** to prove it has been assigned a certain role. Such certificates are given to replicas by the object's owner. They bind the public key of the principal running the replica to the replication role assigned to that replica and are signed with the owner's private key.

5. THE REPLICATION CONTROL PROBLEM

The replication control problem is how to determine which replicas are allowed to propagate state updates and to which replicas state updates should be propagated. In fact, this is access control and reverse access control on a special method - *stateUpdate*. This problem is further complicated by the fact that the various elements of the DSO's state may have different security properties. This refines the granularity of the access control decision which is now also based on the parameters of the *stateUpdate* request - the state change being propagated.

The problem is simplified by the fact that only the replicas of the DSO can exchange *stateUpdate* messages. The DSO's users have no direct access to its state; they can manipulate it only through method requests.

We need to stress an important point here: the problem we are trying to solve is not when a replica should start sending state update messages, or how these state updates should be constructed. That is outside the scope of the security architecture, and is determined by the replication algorithm being used. With replication control, we simply set some data-flow policies that the replication subobject must follow. A replica can send state updates once every second or once every hour, and those state update messages can incorporate every small write or batch a number of such writes, those are all details dependent on the replication protocol. What we want to ensure is that a replica will not send state updates to other replicas that are not trusted enough to even see that part of the object's state in the update. We also want to ensure that a replica will not accept a state update created by another replica which is not trusted enough to modify the part of the object's state in the update. For example, we may want only the replica storing the master copy of the DSO's state to be able to send state updates to the caches. Caches then should be prevented from updating the state of the master replica or of the other caches.

We claim that the replication role set we introduced for the reverse access control problem is also relevant for the replication control problem: consider two replicas in the same role - if they wouldn't have the same rights in sending and receiving *stateUpdate* messages, then they would have different functionality, hence they couldn't be in the same role. Therefore, all replicas with the same role must have the same set of sending and receiving rights for stateUpdate requests.

Now we need to accomodate the requirement that *stateUpdate* exchanges can have different security requirements based on which elements of the DSO's state they modify. This can be done by dividing the DSO's state into disjoint subsets of elements with the same security sensitivity. Those subsets are called **partitions**. State updates for elements in the same partition would then be propagated in the same way though the DSO.

Using these constructs, we can now implement the replication control scheme: for each replication role, we need to specify for which partitions is it allowed to generate any *stateUpdate* messages, and to which roles it can send those messages. This information can be organized in a vector, with one entry for each partition in the DSO's state. Each entry would store the roles to which *stateUpdate* messages for that partition can be sent. Combining the vectors for all the replication roles of a DSO, we obtain the **replication control matrix** for that DSO. This matrix is again stored in the security subobject of each replica.

6. PUTTING THE PIECES TOGETHER

The **access control matrix, reverse access control matrix** and **replication control matrix** fully describe what interactions are allowed between the different replicas and users of a DSO, according to the security policy set by the object owner. These structures are kept in the security subobject, where they are initialized when local objects are created. In order to ensure that the security policy enforced by a local object is the one set by the object owner, these data structures will be digitally signed by the owner with its private key, and sent to the principal creating the new local object together with the user role certificate (and possibly the replication role certificate, if the principal creates a replica).

Replicas and users can interact only after they have authenticated each other. The first step is for the entities to exchange their user/replication role certificates. A user/replica then authenticates itself by proving it has access to the private key corresponding to the public key in its user/replication role certificate (for example by signing a random nonce sent by the other party with its private key).

Once two entities have authenticated each other, they can set up a secure communication channel using some Diffie-Hellman or RSA public key cryptography technique [7]. Secure channels are managed by the security subobject, and provide at least data integrity, authenticity, and possibly confidentiality. The secure channel (identified by a channel ID) is the only thing the security subobject makes visible to the other subobjects. All the details regarding the role assigned to the replica/user at the other end of the channel, keys and cryptographic algorithms used to protect the data on the channel, are managed by the security subobject only, and hidden from the rest of the application.

Before any subobjects part of a local object can perform any actions, they need to check with the security subobject that these actions do not violate the security policies set by the object owner. When a user wants to invoke a method M, it needs to make a call $findReplica(M)$ to its security subobject. This will check the reverse access control matrix, find the replication role allowed to execute M, find a replica in that role (this is done with the help of the Globe Location Service [13], but the details are outside the scope of this paper), establish a secure channel with that replica and return the channel ID for that channel. Finally, the method request is sent on that channel.

A similar scenario happens when a replica receives a request to invoke a method M from one of the secure channels it has established with the users. The replication subobject needs to call $allowAccess(channelID, M)$ to the security subobject. The security subobject uses the $channelID$ to retrieve the role of the user at the other end. It then checks the access control matrix to see if that user role is allowed to invoke M, and returns $true/false$ according to the entry in the matrix.

Finally, replication control decisions are needed when replicas receive *state Update* messages. The first step is to identify the partition to which the state update is targeted. Following that, the replication object needs to call $allowUpdate(channelID, partition)$ to the security subobject, where channelID is the id of the secure channel where the update comes from. The security subobject checks the $channelID$ to retrieve the replication role of the replica at the other end. It then checks the replication control matrix to see if that replication role is allowed to update that partition, and returns $true/false$ according to the entry in the matrix.

7. AN EXAMPLE

In this section we'll show an example on how the security scheme we described in the previous sections can be used to construct a secure Globe application.

Consider a DSO modeling an electronic newspaper: this contains articles and on-line advertising. Those are stored on a set of core replicas. However, we want separation between core replicas dealing with articles, and those dealing with advertising, since the later could be managed by a third party (Doubleclick for example). Newspaper content is pushed by the core group toward a much larger group of caches, which in turn provide this content to the newspaper's readers. Readers fall into two roles: (1) registered users that can only read the headlines, and (2) subscribers who are be allowed to read full articles.

We can model such an application with a DSO that has the following methods: *add_news()*, *add_advert()*, *read_headln()*, *read_article()*. We identify the user roles for this application as the *Editor* (manages articles), *Advertising Manager* (deals with advertising content), *Registered User* and *Subscriber*. *Editors* should not be allowed to add advertising; *Advertising Managers* should not be allowed to add articles; *Registered Users* and *Subscribers* should not be allowed to add anything. Furthermore, *Registered Users* should only be allowed to call *read_headln()*. Figure (3) shows the access control matrix for the object.

Methods

		add_news	add_advert.	read_headln	read_article
User Roles	Editor	True	False	True	True
	Advertising Mngr	False	True	True	True
	Registr. User	False	False	True	False
	Subscriber	False	False	True	True

Figure 3. The Access Control Matrix for the E-Newspaper DSO

As for reverse access control, we identify the replication role set to have three elements - Core Articles Stores, Core Advertising Stores and Caches. Only Caches should be allowed to serve the *read* requests. Only Core Articles Stores should be allowed to execute *add_news()* requests. Only Core Advertising Stores should be allowed to execute *add_advert()* requests. Figure (4) shows the reverse access control matrix for the object.

Finally, for replication control, we separate the DSO's state into two partitions, one for article content (P_0), and the other for advertising content (P_1). Only Core Articles Stores can generate *stateUpdate* messages for P_0, and those messages should be sent only to other Core Articles Stores (active replication) and Caches. Only Core Advertising Stores can generate *stateUpdate* messages for P_1, and those messages should be sent only to other Core Adver-

Methods

Replication Roles		add_news	add_advert.	read_headln	read_article
	Articles Store	True	False	False	False
	Advertising. Store	False	True	False	False
	Cache	False	False	True	True

Figure 4.   The Reverse Access Control Matrix for the E-Newspaper DSO

tising Stores (active replication) and Caches. Figure (5) shows the replication control matrix for the object.

Partitions

Replication Roles		Articles Partition	Advertising Partition
	Articles Store - **ArtS**	ArtS, Ch	Not Allowed
	Advertising. Store - **AdvS**	Not Allowed	AdvS, Ch
	Cache - **Ch**	Not Allowed	Not Allowed

Figure 5.   The Replication Control Matrix for the E-Newspaper DSO

8.   CONCLUSIONS AND FUTURE WORK

In this paper we describe three access control problems we encounter when designing a security architecture for the Globe system. The classic access control problem is extremely general and is found in distributed systems, operating systems and databases security. Reverse access control is common to systems where there is a large number of servers, and users need ways of identifying which of them are the legitimate providers of a given service. Finally, replication control is the problem of adding a security policy to a state consistency protocol.

The general techniques we use to tackle these problems is to organize entities (users, replicas, state elements) into sets with equivalent security properties and to design security policies based on these equivalence sets. This approach is influenced by previous work done on Role Based Access Control (RBAC).

As for future work, we plan to integrate this security architecture in the implementation of the Globe system. We would also like to investigate ways of adding mandatory (site-specific) access control policies to the current architecture. Another topic of research is developing more fine-grained access control mechanisms, based on predicates on environment conditions (time, geographical location, ...) and on the parameters of the method requests.

References

[1] The common object request broker: Architecture and specification. www.omg.org, Oct 2000. Document Formal.

[2] Corba security service specification. www.omg.org, March 2001. Document Formal.

[3] M. Abrams, S. Jajodia, and H. Podell, editors. *Information Security - An Integrated Collection of Essays*. IEEE Computer Society Press, Los Alamitos, CA, 1995.

[4] A. Bakker, M. van Steen, and A. Tanenbaum. From remote objects to physically distributed objects. In *7th IEEE Workshop on Future Trends of Distributed Computing Systems*, pages 47–52, December 1999.

[5] G. Eddon and H. Eddon. *Inside Distibuted COM*. Microsoft Press, Redmond, WA, 1998.

[6] A. Grimsaw and W. Wulf. Legion - a view from 50000 feet. In *Fifth IEEE International Symposium on High Performance Distributed Computing*. IEEE Computer Society Press, Aug 1996.

[7] C. Kaufman, R. Perlman, and M. Speciner. *Network Security*. Prentice Hall, Upper Saddle River, NJ, 1995.

[8] J. Leiwo, C. Hanle, P. Homburg, C. Gamage, and A. Tanenbaum. A security design for a wide-area distributed system. In *Second International Conference Information Security and Cryptology (ICISC'99)*, volume 1787 of *LNCS*, pages 236–256. Springer, 1999.

[9] J. S. Park and R. Sandhu. Rbac on the web by smart certificates. In *ACM Workshop on Role-Based Access Control*, 1999.

[10] C. P. Pfleeger. *Security in Computing*. Prentice Hall, Upper Saddle River, NJ, second edition, 1997.

[11] R. Sandhu, E. Coyne, H. Feinstein, and C. Youman. Role-based access control models. *IEEE Computer*, 29(2):38–48, Febr. 1996.

[12] R. Sandhu and Q. Munawer. How to do discretionary access control using roles. In *ACM Workshop on Role-Based Access Control*, 1998.

[13] M. van Steen, F. Hauck, P. Homburg, and A. Tanenbaum. Locating objects in wide-area systems. *IEEE Communications Magazine*, pages 104–109, January 1998.

[14] M. van Steen, P. Homburg, and A. Tanenbaum. Globe: A wide-area distributed system. *IEEE Concurrency*, pages 70–78, January-March 1999.

THE CORAS APPROACH FOR MODEL-BASED RISK MANAGEMENT APPLIED TO E-COMMERCE DOMAIN

Dimitris Raptis, Theo Dimitrakos, Bjørn Axel Gran, Ketil Stølen
INTRACOM, Greece, drap@intracom.gr: CLRC Rutherford Appleton Laboratory, UK, t.dimitrakos@rl.ac.uk: Institute for Energy Technology, Norway, bjornag@hrp.no: Sintef Telecom & Informatics, Norway, kst@sintef.no

Abstract: The CORAS project develops a practical framework for model-based risk management of security critical systems by exploiting the synthesis of risk analysis methods with semiformal specification methods, supported by an adaptable tool-integration platform. The framework is also accompanied by the CORAS process, which is a systems development process based on the integration of RUP and a standardised security risk management process, and it is supported by an XML-based tool-integration platform. The CORAS framework and process are being validated in extensive user trials in the areas of e-commerce and telemedicine. This paper presents an overview of the CORAS framework, emphasising on the modelling approach followed in the first of the user trials (concerning the authentication mechanism of an e-commerce platform) and it provides some examples of the risk analyses employed in this context.

Key words: security, risk analysis, modelling, e-Commerce.

1. INTRODUCTION

The emerging electronic services (e-services) in the areas of e-commerce, e-health, telemedicine and e-government impose new and increasingly demanding requirements to the underlying infrastructure. A proper understanding of the limitations of the existing infrastructures is an important prerequisite for designing new services with a satisfying degree of security. An improved methodology for risk management is a necessary first step towards verifying and/or improving the security of such systems.

The issues that risk management needs to address are the adequacy of the deployed security mechanisms to meet the application specific security requirements. One of the goals of CORAS is to incorporate suitable risk assessment techniques that address the security requirements of a developed system into appropriate phases of object-oriented software development processes. To this end, systematic model-based risk analysis can help development in two ways:
− Validate that no security aspects are overlooked in a systems' design.
− Provide feedback to refine the security requirements or improve the security mechanisms.

The CORAS project aims to develop an integrated framework for model-based risk analysis of security critical systems. Extensive trials are performed to validate the applicability and effectiveness of the framework in the e-commerce and telemedicine domains. The first trial, using the initial version of CORAS framework, was based on the user authentication mechanism of an e-commerce platform.

CORAS is a European R&D project partially funded by the 5th Framework Programme (FP5) on Information Society Technologies (IST). The CORAS consortium consists of 11 partners from industry, research and academia, from four European countries: three industrial partners, Telenor (NO), Intracom (GR) and Solinet (D), seven research institutes IFE (NO), NR (NO), SINTEF (NO), NCT (NO), RAL (UK), CTI (GR) and FORTH (GR), and an academic partner: QMW college, U. of London (UK).

Section 2 presents an overview of the CORAS model-based risk analysis framework for security critical systems. Section 3 presents some results and experiences of applying the framework on the user authentication mechanism of an e-commerce platform. Conclusions are presented in Section 4.

2. THE CORAS FRAMEWORK

2.1 Objectives

The overall objective for the CORAS project is to provide an integrated methodology to aid the design of secure systems by:
a) Developing a practical framework for a precise, unambiguous and efficient risk analysis, by exploiting the synthesis of risk analysis methods with semiformal specification methods (in particular, methods for object oriented modelling) and computerised tools, in order to improve the risk analysis of security critical systems. The framework will be supported by an adaptable integration platform based on data and information exchange

between risk analysis and modelling tools. The emphasis of this synthesis is on

- adapting an appropriate combination of risk analysis methods to the security critical systems;
- using (semi-)formal modelling techniques in order to isolate the important aspects and obtain an suitable overview of complex security critical systems;

b) Assessing the applicability, usability and efficiency of this framework by extensive experimentation in the fields of e-commerce and telemedicine;

Although the scope of CORAS is security-critical systems in general, it places particular emphasis on information security defined broadly by[1]:

- Confidentiality: ensuring that only appropriate access is allowed to data, both from inside or outside the organisation;
- Integrity: ensuring that no unauthorised changes are made to data – either in storage or transmission;
- Availability: ensuring that data is accessible as required;
- Accountability: ensuring that users are accountable for their security-relevant actions.

2.2   Overview

The main result of CORAS is a tool-supported framework for model-based risk assessment. It is model-based in the sense that it gives detailed recommendations for the use of UML modelling in conjunction with risk assessment. In fact, it employs modelling technology for three main purposes:

1. To describe the target of assessment at the right level of abstraction.
2. As a medium for communication and interaction between different groups of stakeholders involved in risk assessment.
3. To document risk assessment results and the assumptions on which these results depend.

As illustrated Figure 1, the CORAS framework has four main anchor-points.

---

[1] Non-repudiation is also a security concern of cryptographic mechanisms.

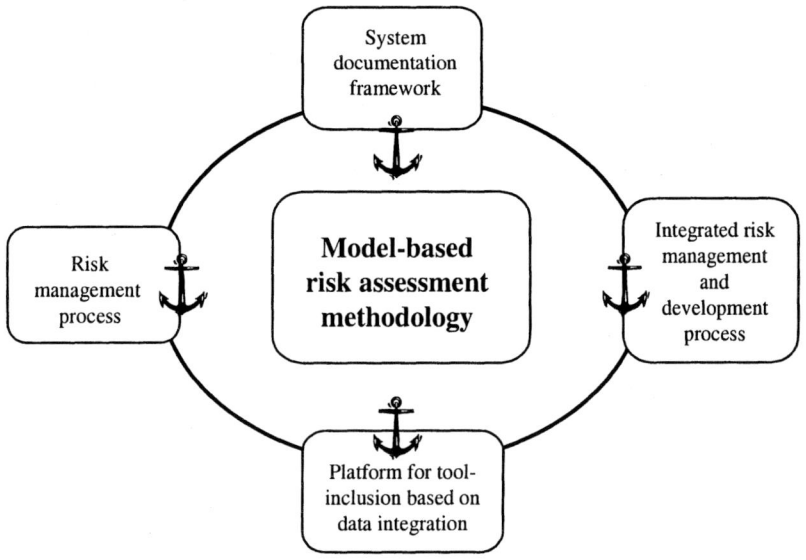

Figure 1. The CORAS framework for model-based risk analysis

The CORAS system documentation framework is based on the ISO/IEC 10746 standard "Basic Reference Model for Open Distributed Processing" (RM-ODP) [8] using UML [12] as modelling notation. RM-ODP defines a reference model for distributed systems architecture, based on object-oriented techniques. The CORAS documentation framework is further elaborated in [4].

The CORAS risk management process is based on ISO 17799 standard "Code of Practise for Information Security Management" [10] as it is complemented by ISO 13335 standard "Guidelines for the Management of IT Security" [9] and Australian standard AS/NZS 4360 "Risk Management" [1].

The CORAS integrated risk management and development process is based on an adaptation of RUP [12] integrating the AS/NZS 4360 risk management process [1] and supporting RM-ODP [8] inspired viewpoint oriented modelling.

The CORAS platform for tool integration will be based on data integration implemented in terms of XML technology. The platform will be built around an internal data representation formalised in XML [15] and in particular XMI [13] for UML models. Standard XML tools will provide much of the basic functionality.

2.3   Risk Assessment

The CORAS risk assessment methodology is build on HazOp analysis [14], Fault Tree Analysis (FTA) [11], Failure Mode and Effect Criticality Analysis (FMECA) [5], as well as CRAMM [2]. Below is a brief description of these methods:

A Hazard and Operability (HazOp) analysis is a systematic study of how deviations from the design specifications in a system can arise, and whether these deviations can result in hazards. The analysis is performed using a set of guidewords and attributes. In general terms, a HazOp analysis is performed as a kind of "brain storming" activity. An analysis team is gathered, consisting of different experts, and headed by a HazOp leader. In its nature, HazOp can make use of almost any kind of information types.

Failure Mode Effect (and Criticality) Analysis (FMEA or FMECA) is an analysis that concentrates on the potential failure modes of individual components. The basis of the FMECA is functional description of the system to be analysed in terms of its components. For each of the components in the system, the aim is to identify all possible or potential modes of failure and classify them according to their criticality. Each potential failure is ranked by the criticality of its effect in order that appropriate corrective actions may be taken to eliminate or control high-risk items. FMECA is usually carried out progressively in two parts. The first part identifies failure modes and their effects. The second part ranks failure modes according to the combination of criticality and the probability of that failure mode occurring. FMECA is a "bottom-up" approach, especially suited to examine all conceivable failure modes and to determine their consequences.

A Fault Tree is a logical diagram, which shows the relation between system failure, i.e. specific undesirable events in the system, and causes that lead to these events. Fault Tree Analysis (FTA) is a method based on deductive logic. First, an undesirable event is defined, and then causal relationships of the failures leading to that event are identified. Fault tree analysis is the most important and most frequently used of the methods available to quantify system performance. It provides not only a mean for system quantification but also a diagram representation of the causes of system failure, which is ideal for communicating the failure relationships.

The British Government's Central Computer and Tele-communications Agency (CCTA) Risk Analysis and Management Methodology (CRAMM), aims to provide a structured and consistent approach to computer security management for all ICT systems. The UK government considers CRAMM to be the standard for the risk analysis of information systems. CRAMM consists of three stages: asset valuation, assessment of threats and vulnerabilities and considering suggested counter-measures. The final tailoring of the security package should include a balance of physical, personnel, procedural and technical security countermeasures. All these decisions should be recorded in the CRAMM software for later review. Where some countermeasures are not implemented this implies that some elements of perceived risk are not covered. The software also allows a series of "what if" questions to be answered with respect to planned changes. In contrast to CORAS however, CRAMM has been angled to support structured systems analysis and design, whereas CORAS aims for object-oriented analysis and design processes. The role of risk analysis for security concerns in systems development is reviewed extensively in [3].

3.   TRIALS

The application of CORAS framework aims at object-oriented systems that are at developing stage but it is equally applicable for developed or maintained systems, like the e-commerce platform.

The e-commerce platform was developed in the context of the R&D project EP-27046-ACTIVE, co-funded by the European Commission under the ESPRIT programme. ACTIVE introduced a generic global Electronic Commerce platform [6] based on Java and the Internet, that supports integrated retail services, providing an intelligent interface upon which the involved players (retailers, suppliers and consumers) can interact. The e-commerce platform used in the CORAS trials constitutes a core part of the ACTIVE system.

In the e-commerce platform, users need be authenticated in order to access the personalised interface or preferences, like shopping lists. Technically, this is not a trivial issue as various alternative approaches have different trade-offs and several implementation pitfalls [7]. In the first trial, the user authentication mechanism used by the e-commerce platform was analysed.

In this section, the modelling approach used to express in UML the user authentication mechanism deployed by the e-commerce platform is presented, and then, examples of the risk analyses performed on the modelled functionality following the CORAS approach are described.

3.1   Modelling

The specification of a system's behaviour can be expressed using UML diagrams like State and Sequence diagrams. In particular, the overall behaviour of a Web application like the e-Commerce platform can be described as a statechart where each state corresponds to a specific HTML page. The exact data presented in the page are abstracted away. Events correspond to links to other HTML pages in the application. The submission of HTML forms and links that pass parameters to the server correspond to events that carry parameters (the form fields, or the link's parameters). Using the same conventions for events (activation of HTML links), specific interactions of users with the application are expressed as sequence diagrams.

In this section the behaviour of the e-Commerce platform with respect to the user-authentication mechanism will be used to present two examples demonstrating this modelling approach.

3.1.1   Dynamic Behaviour Example

Using the modelling approach presented above, the state machine in Figure 2 provides a high level description of the E-Commerce platform behaviour with respect to the user authentication and identification.

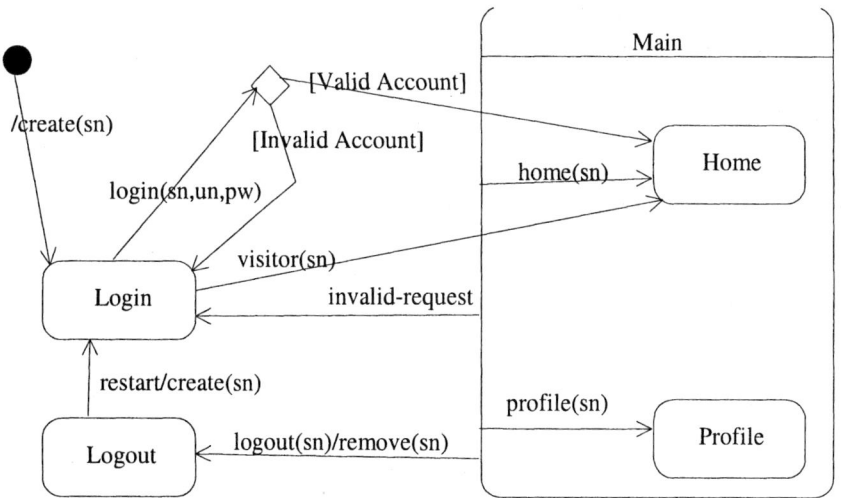

Figure 2. e-Commerce User Authentication behaviour

All HTML pages of the E-Commerce platform (like most web applications) are created using a specific HTML template. The only pages were the template is not applied are the "Login" and "Logout" pages that appear before and after the login and logout. This template contains links to major functions of the application. The superstate "Main" includes the states that can be reached directly from the template. These states will normally include more states corresponding to pages reached from internal links.

The actual behaviour of the E-Commerce platform template has much more states but, for the shake of brevity, only those pertinent to login and registration are shown here:

– The superstate "Main" contains one state for each link in the template of the HTML pages, although only two of them are shown here.

– The state "Profile" is actually a superstate with one state for each HTML page corresponding to categories of user data (i.e., "Registration", "Preferences", "Interests", etc.) but it is included here as a simple state for simplicity because the users can change their password in the initial page.

According to the above diagram, when a user accesses the Login page, the server creates a unique *session ID* to identify the specific client. The session ID is used to associate each user's client with the user's data stored on the server. This session ID is sent to the user's client in all subsequent HTML pages: All HTML links contain the session ID as a parameter. In the state machine above the session ID is denoted as the parameter "(sn)". The login even caries also the username and password as a parameter. Users can also access the platform as visitors without authentication but there are not able to use all functionality like shopping lists.

The server stores the session numbers of all users, both registered (that have been authenticated) and visitors, and associates the registered ones with their corresponding data. A session terminates when a user logs out. In that case, the session number used to identify the session is removed from the database and the user is presented with the "Logout" page.

Invalid requests are those where the parameters were modified with invalid values. In these cases, the server responds with the Login page. Following some time without interaction with the server, a session times-out and the corresponding user is logged off (this holds for visitors as well). The use of time-outs prevents the use of bookmarks for accessing specific pages in the platform.

The use of session numbers for client identification has some repercussions in the behaviour of the application. For example, a session does *not* terminate immediately when a client disconnects. Users can bookmark a specific application page, exit their browser, then restart the browser and go to the specific page via the bookmark before the timeout period expires.

3.1.2   Scenario Description Example

The use of session numbers for client identification can have undesirable consequences if a malicious actor captures a client's session ID. This actor can use the session ID to login in the platform using a second account with an offensive profile (e.g., vulgar names) using the legitimate user's session ID. From that point on, all interactions of the legitimate user with the platform will have the profile of the second account.

This behaviour can be expressed as the sequence diagram in Figure 3.

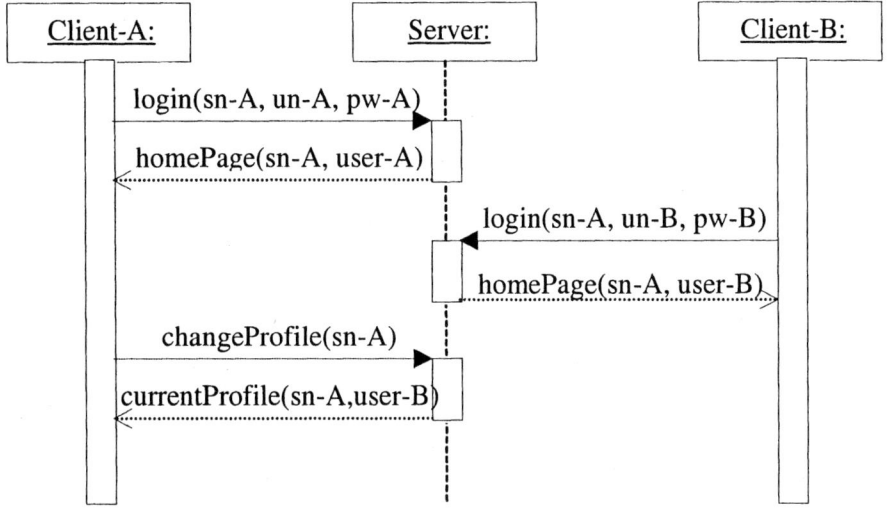

Figure 3. Session hijacking scenario

A legitimate user, Client-A, logs in the platform with the username/password un-A/pw-A, using session ID sn-A, and receives back a page personalised to Client-A user-A. After a malicious user, Client-B, logs in with the session ID of Client-A, the profile that Client-A accesses is the profile of Client-B (e.g., the Client-B's name and shopping lists).

This example is not the result of implementation defects, but a consequence of the design decision to use session IDs for client identification. It should be noted that the use of cookies is subject to similar deficiencies [7].

3.2 Risk Analysis

The risk analysis of the user authentication mechanism deployed by the e-commerce platform was based on models of its behaviour like those presented above. Initially CRAMM was applied in order to provide an identification of assets, which in turn provide a basis and justification for the security requirements that the mechanism need to meet. Then HazOp, FMEA and FTA was performed, examples or which are presented below.

The objective of HazOp, is to identify possible deviations from expected behaviour, as well as their causes and consequences. The expected behaviour was described as a statechart, like the one presented in Figure 2. Using this model, each interaction with the system (event) was systematically analysed. As an example, an excerpt of HazOp applied on a user's request to access the Login page ("/create(sn)" event) is presented in Figure 4 and described below.

No.	Entity	Description	Security attribute	Deviation	Causes	Consequences	Actions	Remarks
1	/create	A user						
1.1	(sn)	requests to	Disclosure					
1.1.1		access the Login Page.		User request captured	Openess of Internet	Not exploitable	N/A	No confidential information transmitted
1.1.2		Server creates a new session		Server response captured	Openness of Internet	SN revealed to capturer	No encryption justified	Deliberate session hijacking is possible
1.2		number (SN)	Manipulation					
1.2.1				A browser or proxy responds with a cached page	Browser or proxy (mis)configuration	User gets a page with invalid SN	N/A	The Login page will returned in the following client request
1.2.2						User gets a SN used by another user	Use large numbers for SN	Inadvertent session hijacking
1.3			Denial / Delay					
1.3.1				User request is blocked by proxy server	Proxy configuration	Server is not accessed	N/A	The server is not accessed
1.3.2				Server response is too slow				Generic deviation
1.4			Unaccountability					
1.4.1				Artificially large number of requests are generated	Deliberate server attack	(1) Creation of too many SNs (2) Server performance degradation	Block access based on client's IP address	Sensitive issue for SN-based user identification

Figure 4. The HazOp table for accessing the Login page

In the HazOp table above, the first column, Entities, correspond to the events of the system behaviour followed by a brief informal description. The Security attributes correspond to possible breach of security requirements (as expressed in Section 2.1) of Confidentiality, Integrity Availability and Accountability. The deviations column presents deviations from normal or expected behaviour, like undesirable (accidental or malicious) interactions with the system. The next columns presents possible causes that enable or cause the deviations, and the consequences of these deviations. The Actions column presents some steps that can be taken to avoid or mitigate the risk of the deviation to occur. Some Remarks are presented in the last column. It should be noted that, in CORAS, there is no standard template for presenting the HazOp results. In particular, the Security attributes, used as guidewords to aid the identification of security violations, may also be customised depending on the aspects of the application that are assessed.

As can be seen from the above table, items 1.1.2 and 1.2.2 can contribute to the generation of the session hijacking scenario presented in section 3.1.2.

FMEA is used to identify possible *failure modes* of individual components. For software systems, like the e-commerce platform, these failures can be wrong results, non-termination, and exceptions or error values returned by function calls to software components. During the trials, the e-Commerce platform was modelled using UML component diagrams. Then each component was analysed, identifying the specific failure modes of the component. Due to the large size of modern software systems the FMEA table may become very large and time consuming to produce. The CORAS trial therefore focused only on small parts of high-level components, like the Web, Application and Database servers of the e-commerce platform. However significant parts of the results were generic and therefore they can be reusable.

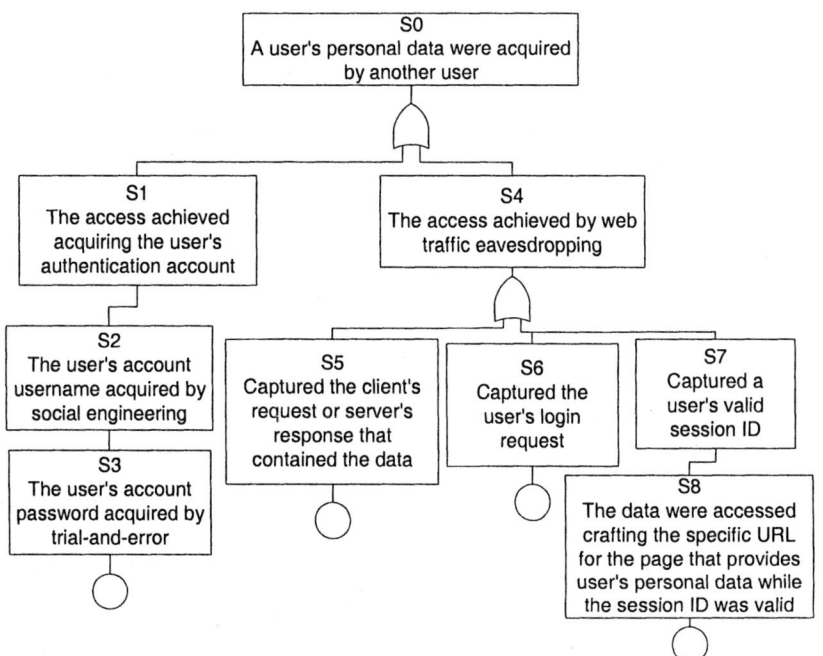

Figure 5. A Fault Tree example for capturing confidential data

The objective of Fault Tree Analysis is to document in a structured way the possible routes that can lead to the violations of security requirements identified by HazOp or failures identified by FMEA. As an example, an excerpt of a Fault Tree demonstrating some possible routes that lead to breach confidentiality by accessing a user's personal data in the e-commerce platform is presented in Figure 5.

The nodes of a fault tree are called *event blocks*, and the root node called *top-event*. The *OR-gates* join alternative means that can lead to their parent nodes. The round circles indicate that the parent events are *basic events* that are not analysed further. In this example, the tree is a branch of larger tree that covers all range of violations of security requirements. There are also more situations resulting in state S0, like circumventing the web server or internal fraud, but these are not presented here.

There is a close relation between the deviations identified by HazOp analysis and the possible component failures identified by FMEA with the fault tree constructed in that these deviations or failures appear as nodes ("event blocks") in the fault tree. For example, item 1.1.2 of HazOp table in Figure 4 identified that a capture of a server response leads to the disclosure of Session ID. This is reflected in FTA tree in *Figure* where state S4 can be achieved by means of reaching state S7. The tighter integration between the complementary risk analysis methods deployed is the following step in the development of the CORAS framework for model-based risk analysis of security critical applications.

4.   CONCLUSION

The CORAS framework for model-based risk analysis of security critical systems offers a structured and systematic approach to identify and assess security issues of ICT systems. The framework is based on modern object-oriented approaches to develop, express and document the systems structure and behaviour.

Extensive trials in the e-commerce and telemedicine domain are performed in order to assess the applicability and effectiveness of the framework and provide insight and feedback for further improvements.

The user authentication mechanism of an e-commerce platform was selected and used for the first trial. Despite the simplicity of the functionality and the wealth of available information on the issue, considerable issues were identified. These issues need be further addressed in the cases where the authentication of users over the Web is critical.

The work presented here is still in progress. Following further developments of the CORAS framework, the secure payment mechanism and the use of agents for automatic transactions with the e-commerce platform will be analysed.

ACKNOWLEDGEMENTS

The authors are grateful to Rune Fredriksen and Eva Skipenes for their valuable comments and corrections. The CORAS project is partially funded by the European Commission under the FP5 Information Society Technologies Programme (IST) by Contract no. IST-2000-25031.

REFERENCES

[1] Australian/New Zealand Standard AS/NZS 4360:1999: *Risk Management*.
[2] Barber, B., Davey, J. *The use of the CCTA risk analysis and management methodology CRAMM*. Proc. MEDINFO92, North Holland, 1589 –1593, 1992.
[3] R. Baskerville, *Information Systems Security Design Methods: Implications for Information Systems Development*, ACM Computing Surveys, Vol. 25, No 4, Dec. 1993, pp. 375-414.
[4] den Braber, F., Dimitrakos, T., Gran, B.A., Stølen K., Aagedal, J.Ø. *Model-based Risk Management using UML and RUP*, Issues and Trends of Information Technology Management in Contemporary Organizations 2002, Information Resources Management Association International Conference, May 2002. (To appear).
[5] Bouti, A., Ait Kadi, D. *A state-of-the-art review of FMEA/FMECA*. International Journal of Reliability, Quality and Safety Engineering 1:515-543, 1994.
[6] EP-27046-ACTIVE, *Final Prototype and User Manual*, D4.2.2, Ver. 2.0, 2001-02-22.
[7] K. Fu, E. Sit, K. Smith and N. Feamster, *Dos and Don't of Client Authentication on the Web*, MIT Technical Report 818, MIT Laboratory for Computer Science, 2001. http://cookies.lcs.mit.edu/webauth:tr.pdf
[8] ISO/IEC 10746 series: 1995 Basic reference model for open distributed processing.
[9] ISO/IEC TR 13335-1:2001: Information technology – Guidelines for the management of IT Security – Part 1: Concepts and models for IT Security.
[10] ISO/IEC 17799: 2000 Information technology – Code of practise for information security management.
[11] IEC 1025: 1990 Fault tree analysis (FTA).
[12] Krutchten, P. *The Rational unified process, an introduction*. Addison-Wesley, 1999.
[12] OMG, *Unified Modeling Language (UML) Specification*, Ver. 1.3, Mar. 2000.
[13] OMG, *XML Metadata Interchange (XMI) Specification*, Ver. 1.1, Nov. 2000.
[14] Redmill, F., Chudleigh, M., Catmur, J. *Hazop and Software Hazop*. Wiley, 1999.
[15] World Wide Web Consortium, *Extensible Markup Language (XML) v1.0*, W3C Recommendation, Second Edition, 6 Oct. 2000.

TOWARDS SECURITY ARCHITECTURE FOR FUTURE ACTIVE IP NETWORKS

Dusan Gabrijelčič
and Arso Savanović
and Borka Jerman Blažič
Institute Jožef Stefan
Ljubljana, Slovenija
dusan@e5.ijs.si, arso@e5.ijs.si, borka@e5.ijs.si

Abstract   Active networks allow user-controlled network programmability. A security framework has to assure that an active networks infrastructure will behave as expected and will efficiently deal with malicious attacks, unathorized attempts to execute active code etc. We present here a security architecture that is designed within the FAIN project and aims at supporting multiple heterogeneous execution environments. We argue for the pros and cons as well as why we have selected the specific components and also take a look at their interworking in order to provide the security services to the execution environments our active network node hosts.

Keywords: Active Networks, Security Architecture, Active Packets, Security Management

1.   INTRODUCTION

Active networks enable their users to program network elements by injecting the code in the network which provides new services for the users in the network. Such flexibility raises many security concerns. Security architecture as described in the following paper, dealing with these concerns, is designed and developed in a FAIN[2] project.

FAIN project goal is to provide open, flexible, high-performance and secure active network infrastructure. Such infrastructure consist of interconnected active and passive nodes. Active nodes provide support for various active networking technologies and are able to support various active networking applications in control, management and data plane. Core of FAIN active network (AN) is active network node (ANN).

2.   FAIN ACTIVE NODES

The FAIN Reference Architecture consists of Active Applications (AA), Virtual Environments (VE), Execution Environments (EE) and Node Operating System (NodeOS).

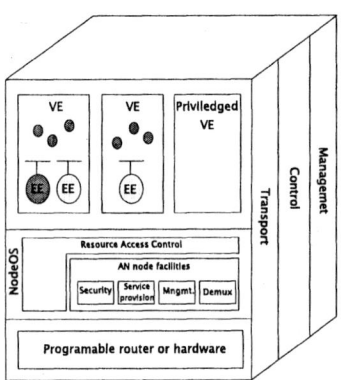

Figure 1.   FAIN Active Node reference architecture

Active Applications/Services are active code executed or running in Active Nodes. Active code enable, support or enhance communication between active network users.

Execution Environments are environments in the active node where the active code is executed. Execution environments represent programming model that active code comply to and abstractions on which Active Applications can be build. A privileged EE in privileged Virtual environment (PVE) manages and controls the Active Node and it provides the environment where network policies are executed. Multiple and different types of EE are envisaged in FAIN. EEs are classified into Virtual Environments (VEs), where services can be built and interact with each other. VEs represent principals on the node. Interconnected VEs form a truly virtual network.

NodeOS is an active node operating system and provides basic abstractions for network communication, resources and inter EE communication. It manages node resources which are accessible through NodeOS APIs. NodeOS functionality is divided into several subareas, which are described in the following paragraphs with short emphasis also on their role in overall node security:

Demultiplexing/Multiplexing subsystem. It filters, classifies and diverts active or passive packets. Flows of packets arrive at the node and they should be delivered to the VE and consequently to the service inside the VE they are destined for. Demultiplexing represents the entry point of the node for packets, which can be seen as entities that request processing on the node and will use node resources. At this point node is protecting its own resources. In multiplexing subsystem packets are sent towards next hops; here, well behaved active node should protect other nodes.

Resource Control Framework (RCF). Through resource control resource partitioning and access to the resources is provided. VEs are guaranteed that consumption stays within the agreed contract during an admission control phase static or dynamic. It provides also guaranteed share of resources to the node subsystems and privileged virtual environment.

Active Service Provisioning (ASP). As flexibility is one of the requirements for programmable networks partly realised as service deployment either on the fly or static, the NodeOS must support it. ASP must work closely with Security subsystem in the way that only code and services that meets certain security criteria can be made available on the node.

Management subsystem. Management subsystem provides interfaces that are meaningful on the NodeOS level. Among them there are interfaces that enable the management system to insert security policies and register entities as principals on the active node.

Security subsystem. Security subsystem is a collection of needed security services, mechanisms and elements that are core of the security architecture. Together with other subareas of the node this subsystem has to fulfill the security goals as stated in the next section.

3.   FAIN SECURITY ARCHITECTURE: GOALS AND SCOPE

FAIN general active networking model follows the general active networking model as described in [3]: the primary unit of communication is a packet, primary goal is communication and not computation in the network, ANN are interconnected with various packet-forwarding technologies, AN consist of multiple domains controlled by different administrations and end systems and intermediate systems have same architectural components. To this common starting point we can add the following general AN properties: not all nodes in the network need to be active, packets used for communication can be both active and "passive", AN can provide per user service in the network by injection of the code in the network both in-band or out-of-band, active application can keep state in the network or in the packets, active packets can legally change in the network, there is no unified address space in AN and packet resource consumption in the ANN and AN is hard to predict and control.

All AN properties as stated, have serious security consequences on AN and ANN operation. They raise many security related questions: where the packet has come from, is it the same as when sent from data origin, who has send the packet, can the code that packet address/carry be safely run on the node, does it access privileged node resources and interfaces, how many resources can/will the packet consume on the node etc. To tackle these questions we have identified a set of possible security architecture goals. From this set we have focused in initial phase only on few, high priority ones, that protect the AN infrastructure from intentional or unintentional misbehaving of AN users. Our primary security architecture goals are authentication, authorization, policy enforcement, active code and packet integrity, code verification and audit.

Because FIAN active node supports multiple and different execution environments and technologies we have designed and focused on security architecture on NodeOS level as common denominator of the node.

3.1.   Authentication, authorization and policy enforcement

FAIN ANN is essentially a multi-user computing system. As in any such system, enforcement of access control is a requirement of high significance within every FAIN ANN. On the other hand, FAIN aims at developing a flexible system. In order to achieve the desired level of granularity and flexibility we decompose access control into authentication, authorization and policy enforcement.

Authentication is a process of verifying an identity claimed by or for a system entity. Authentication is a basic security service that other services depend on. From a range of solutions based on either symmetric or asymmetric cryptography we have chosen to provide authentication service on the basis of digital signature mechanism. The main reason for this decision is that the active packet can pass many nodes in the network where its data origin is authenticated. Proper authentication solutions that require handshake between active packet source and every possible active node in the packet path are too costly, even though they can be amortized over time. Using digital signature mechnisms for authentication requires, that every user as well as every active node has at least one valid key pair, and existence of a public key infrastructure.

Authorization is a process of authorizing user access to certain node resources. Resources can be hardware like CPU, memory, storage and link bandwidth or functional like special purpose files, routing tables, policy and credentials entries and databases etc. All node resources are accessible through set of certain node interfaces. Authorization decision is made based on information about subject requesting a resource, object being accessed, action requested and possible environment of the request. To support such decision relevant security policies must be available in the node policy database and the security contexts of the subject and the object have to be built and kept unforgeable on the node. Authorization decision is then enforced in node enforcement engines.

Policy enforcement is a task of many node enforcement engines which resides in the node subsystems. The task is twofold: upon request to access certain node interface the engine suspends the request, collects the relevant information (object, subject, action and environment) and asks for authorization decision. When authorization decision is made, the enforcement engine either discards or resumes the request. In both cases the resulted decision and enforcement engine action can be audited.

3.2.   Active code/packet integrity

Active code and packet integrity is an important issue and thus high priority goal in FAIN AN. In both in-band (code and data in packet) and out-of-band (code is provided separately) active network approaches the content of the active packet has high impact on the active application. Every change of the packet like replacing or modifying its content, possible cut and paste or replay attacks can have unpredictable consequence of the behavior of the active application as a whole. Providing integrity service for active packet traversing many nodes is also an issue. Active packets can carry a state or can change (packet or in-band code) in the network. This means that the data integrity calculated at the originating node can be verified only if the data that the integrity service covers has not changed in between the source and the point of its verification. Logical consequence of the stated is, that the integrity service has to be provided separately for the variable and static data in the packet. Packet integrity is closely related to the authentication as presented in section 3.1. There can be no direct authentication if the integrity property of the data which origin is being authenticated is not provided as well. Digital signature used for authentication provides also integrity property for the data but only for the static parts of the packet which the digital signature covers. To protect the packet's variable content when the packet traverses the network we use a keyed hash as protected integrity token and sequences for replay attack prevention in between pair of neighbor active nodes. Such integrity service requires shared secret between a pair of neighbor nodes; this secret can be established either via management system or dynamically with suitable key exchange protocol. It has to be noted that such integrity protection protects data only between two neighbor active nodes; trust in integrity of data that has passed many nodes means that we trust all the nodes, the packet has already passed, which can be a weak assumption from security point of view.

The last observation can be specially true for the active code if in-band approach is used. In this case data and code can be mixed together and the "code" can change on every hop. Integrity of such code can be protected only per hop and not between source and destination.

3.3.　　Code verification

Protecting the active code integrity is a first step to ensure non- modification of the transient code. However this is considered pretty basic and we need to go beyond that in order to achieve higher level of security. The active code has to be somehow marked and tightly coupled with one or more entities, based on which further security decisions can be made. As an alternative, code in conjunction with EE has to exhibit certain properties which can be verified on the node.

Verification can enable us to trust to some extent that the active code will behave safely and properly and that we can have some guarantees on its resource usage on the node and in the network. But we shall say in general that verification provides only enhanced trust in proper and safe code execution, which is usually not related to the trust in the entity on behalf of which the code is executing. Besides the root of trust is in many cases different for the code and the packet that triggers the code on the nodes. Code verification can help an ANN decide whether to run the newly received code. If the code fails the verification test, it is not trusted and it is dropped or alternatively it can run in an EE with minimal facilities available. In the latter case the EE is the same one that will be used to run anonymous active code. Broadly we can divide verification in two groups: first, in which we trust in an entity in the system that enable us to believe in some code properties and second, belief in code properties by design (safe languages or/and interpreters) or additional means like proof carrying code [7]. In the first, general case, node can trust user, network owner, manufacturer, code packager, code repository or dedicated organization, which can in various ways guarantee some code properties. Code can be source inspected, tested or proved otherwise to work as claimed. To enable trust in an entity we use digital signature mechanisms and code certificate, in which the claimed properties or conditions of use of the code can be stated and verified prior the code installation/usage directly on the NodeOS level. In the second case we have to trust EEs to properly verify the code prior and/or during its execution.

3.4. Audit

The Audit Manager component is an integrated part of the security architecture. Via this component all events occurring from the usage of the security subsystem are implicitly logged for further future usage. It also provides an interface to explicitly log any other events coming from other parts of the FAIN architecture in a clear and homogeneous way. Modern computer systems do not emphasize enough on the significance of the audit facilities. However audit tools help in realizing possible security leaks (or even preventing some) and make sure that mistakes are not repeated.

4. FAIN SECURITY ARCHITECTURE

Figure 2 depicts a FAIN active network node with basic security architecture components. As depicted, FAIN security architecture roughly comprises three parts: security subsystem, other ANN security components, and external security support facilities. Note that the scope of initial FAIN security architecture does not include EE layer of FAIN ANN architecture.

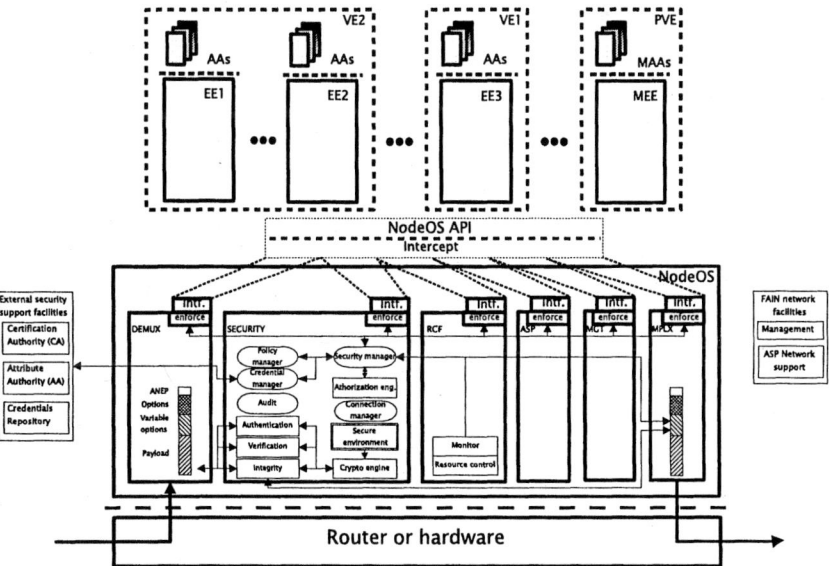

Figure 2. FAIN security architecture

4.1.   Security subsystem

Most of security critical decisions are made by security subsystem, which is one of several subsystems within an ANN. The Security subsystem is also responsible for management of security critical data, such as encryption keys, credentials, and policies.

This subsystem is the core of FAIN security architecture and includes the following components:

Crypto Engine: performs the actual cryptographic operations, such as symmetric encryption/decryption, asymmetric encryption/decryption, and hashing. It implements various cryptographic algorithms, which are used by other components in the security subsystem.

Secure Environment (SE): in a secure fashion stores various encryption keys, which are required by crypto engine. For example, SE stores ANN's public key pair (private and public key) and all symmetric keys that an ANN shares with its AN neighbors (one per neighbor).

Connection Manager: is used to manage secure associations with neighbor AN nodes. Associations can be configured manually or their configuration can be supported by automatic management and by triggering a key exchange protocol with neighboring ANN.

Integrity Engine: checks the integrity of active packets and active code. It depends on integrity protection data contained within an active packet and on crypto engine to do the necessary cryptographic operations.

Verification Engine: performs code verification (at NodeOS level), if any. It may depend on special data contained within an active packet and on crypto engine to do the necessary cryptographic operations.

Authentication Engine: verifies the authenticity of active packets. It depends on authentication data contained within an active packet and on crypto engine to do the necessary cryptographic operations.

Enforcement engines: which intercept requests to NodeOS API interfaces and pass them with request specific information to security manager. Security manager responds with authorization decision and enforcement engines enforce this decision. They are distributed and not present only in security subsystem.

Security Manager: accepts request from enforcement engines, collect request related credential information, accessed object policy information and ask authorization engine for authorization decision. This decision is passed back to enforcement engines. Provides NodeOS interfaces for managing AN users related credentials and security policies. It can be used also for caching of the authorization decisions.

Authorization Engine: is responsible for making a decision whether a given user request to execute specific action or to access/manipulate particular object within an ANN is authorized or not. Authorization engine provides this "service" to all enforcement engines in an ANN.

Policy Manager: when asked by the security manager, searches policy DB and returns all security policies, that are relevant for a particular request, which is currently subject to authorization. It also provides facilities for editing entries in policy DB, either manually by an authorized user, or automatically, i.e. download policies from a centralized policy server.

Credential Manager: when asked by authorization engine, searches credential DB and returns all credentials, that are relevant for a particular request, which is currently subject to authorization. It also provides facilities for editing credential database, either manually by an authorized user, or automatically, i.e. search and download credentials from an external credential repository.

Audit: will be the place where all security architecture's components audit their function in order to be used later in resolution of problems or even to make decisions. E.g. an Intrusion Detection System would use a view of the audit DB in order to recognize attacks against the system. The audit could be also distributed for survivability reasons.

5. TOWARDS IMPLEMENTATION

The basic scenario we had in mind developing security architecture was an active packet passing one or more active nodes and triggering execution on these nodes. While executing or processing the packet, the active code on the nodes can access privileged NodeOS interfaces.

To enable security architecture operation as described in section 3 and to support basic scenario, we have chosen the ANEP packet header [6] as a carrier of needed information for integrity and access control. Basic ANEP options that support integrity, authentication and basic resource usage control are shown in the figure 3. Hop by hop option (3b) consist of KeyId, 64 bit value uniquely identifying security association, 64 bit sequence field and HMAC field holding keyed hash as defined in RFC2104 [11]. Keyed hash covers entire ANEP packet except keyed hash value itself. At the moment security associations are setup via management system; we are developing set of protocols that will support dynamic secret key exchange, sequence synchronization and discovering of the first hop active node.

The credential option (3c) was designed as simple four-tuple option; there can be multiple such options in the packet for additional flexibility.

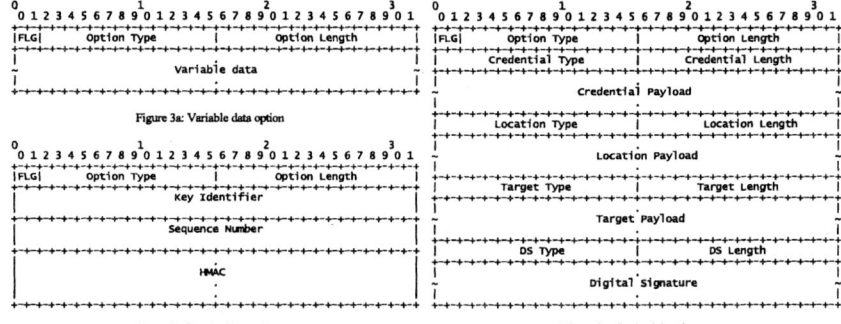

Figure 3.   ANEP packet options

Credential as designed can be X.509 certificate, X.509 attribute certificate [1] or Keynote credential [8]. Each credential option is validated and signatures verified prior the packet triggers execution on the node and the target field in this option help node to use credentials only in administrative domains, nodes or node subsystems as intended by packet sender or administrative domain boundaries controllers.

Variable option (3a) is meant as a place where active applications can store variable data and is a consequence of integrity discussion in the section 3.2. While is hard to adapt existing approaches to support such option, newly developed will support it out of the box.

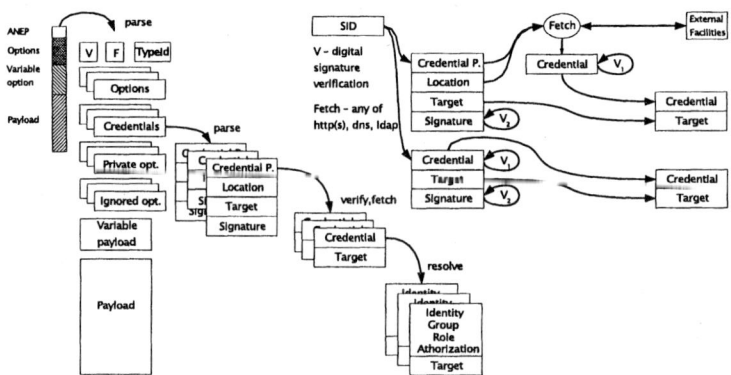

Figure 4.   Active packet, identity/role resolving, credentials validation and signature verification

The figure 4 shows the basic active packet already processed on the active node (left) and process of verification of the one or more attached credentials (right). Credentials can be carried in the option or the option carries the location of the credential; in the second case, the credential has to be obtained from remote repository. In both cases we validate the credential and the possible certification path. With the public key obtained from validated credential, digital signature of the static part of the packet is verified. The next step is resolving identity, group, role or authorization information from the credential. This information is then used in the authorization decisions while the packet (code) access the domain or privileged node interfaces. Domain on the node is a set of related processes/threads that perform certain service for or on behalf of the entity.

The crucial issue on the node regarding access control is security context of the request (packet) passing the node. While the context on the node is defined by entity properties (identity, role or group) and domain information, such context has to be transfered between nodes. While the figure 4 does show how entity information can be transfered through the network, it doesn't show domain related part. Domain is hidden in VE identifier and active code related identifiers. These identifiers must match the domain the packet is destined for.

Policies used in the process of the providing authorization decision are set up on the node by the FAIN management system.

6. RELATED WORK

FAIN aims to develop a heterogeneous ANN, allowing coexistence of various technologies that enable installation and execution of active code within an ANN. Consequently, FAIN security architecture is aimed at providing a more general solution which provides necessary protections for such an heterogeneous system. This is reflected by the fact that security architecture we have presented does not incorporate details of specific EEs that exist in the FAIN ANN. Its goal is to be as EE independent as possible and provide a common set of basic security services required by all AN enabling technologies. Some research projects on active networks have already tried to tackle the issue of security [5, 12]. Contrary to FAIN, all these approaches are tied to specifics of particular model of programmability. When designing a more general AN security architecture, which is the case in FAIN, these specifics can not be assumed. Java Security Architecture [9] proved to be useful for AN security, but again it is technology specific and it also has some drawbacks [10]. There has also been some more general work on AN security [4] which has also working specific implementation [13]. But the general work is still in draft stage.

7.   CONCLUSIONS

We have presented in this paper a security architecture for future IP active networks as it is done in the context of FAIN project. We try to tackle the high priority security requirements such as authentication, authorization, policy enforcement, active code and active packet integrity and verification and last but not least audit. We have analysed the main design decisions that we have taken and the reasons why we decided to follow them. Subsequently we have presented the components of a security architecture that will be used by multiple heterogeneous execution environments within the same active node. We also provide a look in the interworkings of the architecture and its decision-making logic. A prototype implementation of the presented active network security architecture is currently under development, which will be used for exploring the advantages and drawbacks of our approach.

8.   ACKNOWLEDGEMENTS

This paper describes work undertaken and in progress in the context of the FAIN - IST 10561, a 3 year project during 2000-2003. The IST program is partially funded by the Commission of the European Union. The FAIN consortium consists of University College London - UK, Jozef Stefan Institute - Slovenia, NTUA - Greece, Universitat Politecnica de Catalunya - Spain, Deutsche Telekom Berkom - Germany, France Telecom/CNET - France, KPN - The Netherlands, Hitachi Europe Ltd. - UK, Hitachi, Ltd. - Japan, Siemens AG - Germany, ETH - Switzerland, Fraunhofer FOKUS - Germany, IKV++ GmbH - Germany, INTER-GAsys - Spain, University of Pennsylvania - USA.

References

[1] ITU-T X.509 (2000) — ISO/IEC 9594-8:2000 - Information technology - open systems interconnection -the directory: Public-key and attribute certificate frameworks. Final Draft International Standard, June 2000.

[2] Fain project home page, *http://face.ee.ucl.ac.uk/fain/*.

[3] Active Network Working Group. Architectural Framework for Active Networks Version 1.1, december 2001.

[4] Active Networks Security Working Group. Security Architecture for Active Nets, maj 2001.

[5] D. Scott Alexander, William A. Arbaugh, Angelos D. Keromytis, and Jonathan M. Smith. Security in active networks. In *Secure Internet Program-*

ming: Issues in Distributed and Mobile Object Systems, Lecture Notes in Computer Science State-of-the-Art. Springer-Verlag, 2000.

[6] D. Scott Alexander, Bob Braden, Carl A. Gunter, Alden W. Jackson, Angelos D. Keromytis, Gary J. Minden, and David Wetherall. Active network encapsulation protocol (anep). Active Network Group draft, july 1997.

[7] Andrew W. Appel, Edward W. Felten, and Zhong Shao. Scaling proof-carrying codeto production compilers and security policies. whitepaper, January 1999.

[8] Matt Blaze, Joan Feigenbaum, John Ioannidis, and Angelos D. Keromytis. RFC 2704: The KeyNote trust-management system, version 2, september 1999.

[9] Li Gong. Java security architecture (JDK1.2). Technical report, Sun Microsystems, oktober 1998.

[10] Active Networks Working Group. SANTS Security Overview, May 2000.

[11] H. Krawczyk, M. Bellare, and R. Canetti. Hmac: Keyed-hashing for message authentication. RFC2104, Informational, februar 1997.

[12] Zhaoyu Liu, Prasad Naldurg, Seung Yi, Roy H. Campbell, and M. Dennis Mickunas. Seraphim: Dynamic interoperable security architecture for active networks. In *Proceedings OpenArch 2000*. University of Illinois, Urbana-Champagain, marec 2000.

[13] Murphy S., Lewis E., Puga R., Watson R., and Yee R. Strong security for active networks. In *IEEE OPENARCH 2001 Proceedings*, april 2001.

COMBINED FINGERPRINTING ATTACKS AGAINST DIGITAL AUDIO WATERMARKING: METHODS, RESULTS AND SOLUTIONS

Martin Steinebach, Jana Dittmann, Eva Saar
Fraunhofer IPSI Germany, martin.steinebach@ipsi.fhg.de
Platanista Germany, HTWK Leipzig University of Applied Sciences,
jana.dittmann@platanita.de
T-Systems, Germany, Eva.Saar@t-systems.de

Abstract: While a reliable protection against illegal copies does not exists today, tracking of illegal copies and prove of ownership are important detection functions, which can be realized by using passive security mechanisms of digital watermarking. Recent research has identified many watermarking algorithms for all common media types ranging from printed matter to multimedia files. The main topics of interest concentrates on transparency and robustness: The watermark must not reduce media quality and should be detectable after most common media operations and attacks. Algorithm security is discussed with regards only to key space most of the times, while especially for customer identification known as active fingerprinting specialized attacks like coalition attacks are known. Digital fingerprinting raises the additional problem that we produce different copies for each customer. Attackers can compare several fingerprinted copies to find and destroy the embedded identification string by altering the data in those places where a difference was detected. Few approaches have been introduced for image and video watermarking schemes, but there are no observations for audio fingerprinting techniques. In our paper we discuss methods for secure customer identification by digital fingerprinting for audio data. We describe first two algorithms by Boneh et al. [BoSh95] and Schwenk et al. [DBS+99] and then combine these schemes with an audio watermarking algorithm for practical evaluation of their coalition resistance to detect illegal copies. We provide test results and evaluate the security against different types of coalition attacks.

Key words: watermarking, customer authentication, fingerprinting algorithms, coalition attacks

1.   BACKGROUND AND MOTIVATION

The expansion of digital networks all over the world allows extensive access on, and reuse of, visual material. Problems include unauthorised taping, reading, manipulating or removing of data, which might lead to financial loss or legal problems of the producers and creators. Thus, designers, producers and publishers of digital data like images, video, audio or multimedia material are seeking technical solutions to the problems associated with copyright protection of multimedia data. Therefore the Internet has become in many cases a trading place for illegal copies of movies, music and software. Thus, systems are required which provide environments where digital data can be signed by authors or producers as their intellectual property, i.e. by embedding private or public information into the video data, to ensure and proof ownership rights on the produced video and audio material during its distribution. Digital watermarking in combination with active fingerprinting algorithms offers a solution to trace illegal copies. We introduce now the essential watermarking parameter for authentication and show customer-tracking concepts. In chapter two we discuss our general design of an collusion resistant audio fingerprint watermarking algorithms by using two general fingerprint construction schemes from [BoSh99] and [DBS+01]. In chapter three we introduce our test environment and the tested coalition attacks. Furthermore we describe the test procedure, the test tool and our test results with our implementation. We summarize our paper with a conclusion and directions of further work.

1.1   Customer authentication watermarking

For tracking the source of the copy an identification method is required. The copies must be personalized to decide who is responsible for the copyright violation. The goal of active digital fingerprinting is to embed customers IDs in the digital media. Personalization is of course only possible if the customer or user can be identified like in web shop environments, copies of e.g. songs ripped from CDs bought in stores cannot be marked this way. Embedding a customer ID brings along a number of requirements to the watermarking algorithms.

- Transparency is a common requirement for marking digital media in e-commerce environments as the quality of the content acting as a cover for the watermark must not be reduced.
- Robustness is necessary against common media operations like lossy compression and format change.

- Payload must be high enough to include the customer ID. This can become a critical requirement as we see in section [2]. In our paper we do not focus on that aspect in detail.
- Security is of special importance in this case as the existence of several copies of the same cover with different embedded customer Ids enables a number of specialized attacks commonly called coalition attacks. In our paper we want to evaluate this parameter in more detail for coalition attacks.
- Complexity has to be low enough to enable online and real time marking. A customer who wants to download a song is not willing to stay online for a long time until his personalized copy is available. As there will be multiple customers at the same time and media data may have a playing time of an hour or more, either streaming concepts or multiple real time embedding speed will be necessary.
- Verification should be performed in a secret environment. The shop owner holds a secret watermarking key to embed and read the watermarks. Customers do not know this key as attackers could easily verify their success with it.

1.2   Tracking of illegal copies

With the concept of customer authentication watermarking tracking of illegal copies becomes possible. It is common to have a user ID in web shops for services like customer accounts, tracking of orders and easy login. This ID can be used either directly or indirectly for personalization of digital data.

The direct method uses a watermarking algorithm to embed the ID in the downloaded content. For added security, the ID may be encrypted together with extracted content features to disable copy attacks. The indirect method uses a database. The download of a specific media are given a serial number consisting of a media ID and a running number. When a customer downloads a file, the actual serial number is embedded and stored in the database together with the customer ID. This allows tracking the copy without personalizing it. It would also allow to create pre-marked copies and to store them for use in times of high activity.

Both methods are used in the same way regarding the security framework. They are embedded when a user who authenticated himself at the time he logged in the web shop starts to download some digital content. Based on the user ID, the seed of the personalization process is as secure as the shop system.

Now every user receives an individual copy of the digital content and is responsible for guarantee legal distribution. A secure distribution channel may be necessary to ensure the individual copy is not duplicated while web delivery. E.g. a session key protocol could be applied based on a PKI framework also used for user authentication.

If a copy of the web shop offers is found in Internet trading places, on illegal CDR copies or similar places, the shop owner can use his secret key to retrieve the watermark and identify the origin of the illegal copy. This requires specialized search engines or equal mechanisms to track copies in the Internet. Without a way to find illegal copies, the watermark-based security mechanism cannot help to track illegal usage. An additional watermark with an owner ID could help to identify content coming from the specific web shop in combination with firewall concepts like the ones we describe in [DKL+02]. As an alternative, audio hashing concepts together with search engines and databases of all content sold by the web shop could be applied.

2.   DESIGN OF FINGERPRINT WATERMARKING

To solve the problem of the coalition attack, we use the Boneh-Shaw fingerprint and the Schwenk-Ueberberg fingerprint algorithm [BoSh95], [DBS+99]. Both algorithms offer the possibility to find the customers, which have committed the coalition attack. In our last work in [DEV+01] we have introduced for both schemes a video fingerprinting solution and the coalition resistance. To mark the video, we generate positions within the frame to embed the watermark information (in the video the positions stand for scenes). Each customer has his own fingerprint, which contains a number of "1" and "0". Each fingerprint vector is assigned to marking positions in the document to prevent the coalition attack. The only marking positions the pirates cannot detect are those positions, which contain the same letter in all the compared documents. We call the set of these marking positions the intersection of the different fingerprints. In the following two subchapters we summarize the two fingerprinting schemes which are used and show how we apply these schemes to digital audio watermarking.

2.1   Fingerprinting concepts

A digital fingerprinting scheme consists of a number of marking positions in the document, a watermarking algorithm to embed letters from a certain alphabet at the marking positions, a fingerprinting algorithm which selects the letters to be embedded for each marking position depending on the number i of the copy and a pirate tracing algorithm which, on input of a modified document, outputs at least one number i of a copy that was used in constructing the modified document.

2.1.1 Schwenk Fingerprint Scheme

The Schwenk et al. approach [DBS+99] approach puts the information to trace the pirates into the intersection of up to d fingerprints. In the best case (e.g. automated attacks like computing the average of fingerprinted images) this allows us to detect all pirates, in the worst case (removal of individually selected marks) we can detect the pirates with a negligibly small one-sided error probability, i.e. we will never accuse innocent customers.

The fingerprint vector is spread over the marking positions. The marking positions for each customer are the same in every customer copy and the intersection of different fingerprints can therefore not be detected. With the remaining marked points, the intersection of all used copies, it is possible to follow up all customers, which have worked together. Another important parameter is the number n of copies that can be generated with such a scheme. The scheme uses techniques from finite projective geometry [Hir98], [BeRo98] to construct d-detecting fingerprinting schemes with q+1 possible copies. These scheme needs $n=q^d+q^{d-1}+...+q+1$ marking positions in the document. As we see this can be a huge length and can cause problems with the capacity of the watermarking scheme. The idea to built the customer vector is based on finite geometries and the detailed mathematical background ´can be found in[DBS+99].

2.1.2 Boneh-Shaw Fingerprint Scheme

The scheme of Boneh and Shaw [BoSh95] was designed to recognize coalition attacks, with a different approach. The method generates fingerprints for each customer containing a different number of zeros and ones. After the coalition attack the detection function we do not necessarily find all pirates. Furthermore with a (any arbitrary small) probability ε we get the wrong customer based on the different number of zeros in the detected fingerprint.

The number of customers is q and with q and ε we calculate the repeats d. The fingerprint vector consists of (q-1) blocks of the length d ("d-blocks"), the total length of the embedded fingerprint computes as $d*(q-1)$. Depending on the repeats the customer vector can be very long and cause problems with the capacity of the watermarking algorithm. The idea to built the fingerprinting vector for each customer is simple: The first customer has the value one in all marked points, for the second customer all marked points without the first "d-block" are ones, in the third all marked points without the first two "d-blocks" are ones etc. The last customer has the value 0 in all marked points. With a permutation of the fingerprint vector we get a higher security, because the pirates can find differences between the copies, but they can't assign it to a special d-block. Detailed mathematical background can be found in [BoSh95].

2.2 Fingerprinting algorithm combined with audio watermarking

To embed the customer information generated by the fingerprinting algorithm to trace illegal audio copies we use a digital watermarking algorithm. Current digital watermarking techniques usually would embed the generated fingerprinting information FP randomly all over the audio sequences with the disadvantage, that the intersection of the proposed fingerprints cannot be used to find attackers after comparing attacks of different customer copies. To use the excellent properties of the fingerprint to conclude to the customers which attacked the watermark we build a watermarking scheme as introduced for image and video data in [DBS+99] and [DHV+01] with a fixed number of **marking positions** in each copy of the audio. The **fingerprinting algorithm** selects the letters, the FP vector over the binary alphabet {0,1}. The **watermarking algorithm** embeds this binary FP vector at the chosen marking positions.

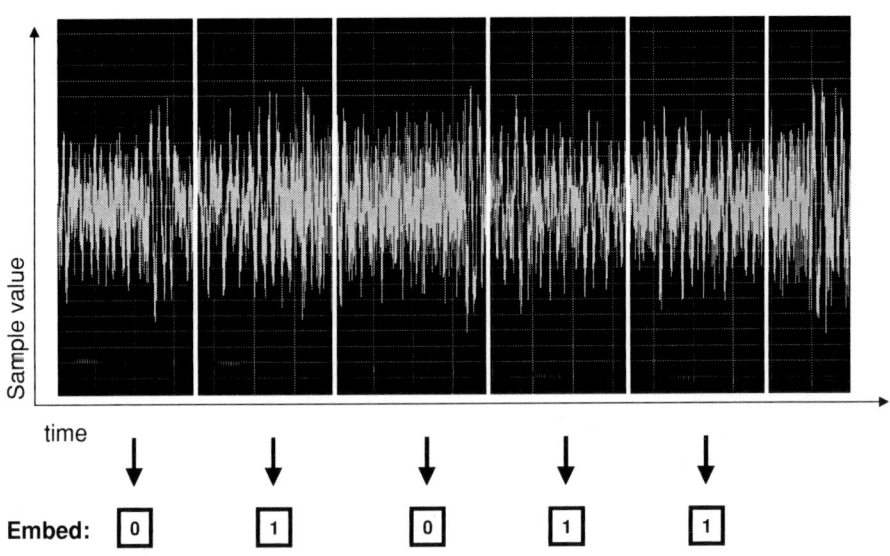

Figure 1: Audio watermarking over time

Generally watermarking algorithms use different methods to embed a message M into a cover C. In this paper, we use an audio watermarking algorithm to embed a fingerprint bit vector FP as M. The way M is embedded in C is relevant for the security of the watermarking and fingerprinting combination: An PCM audio stream consists of a sequence of audio samples over time. Most multi-bit message audio watermarking algorithms use a group of successive samples, e.g. 2048, to embed a single bit of the complete message. Figure 1 illustrates this: The bit sequence 01011 is embedded in a 1 second audio segment by separating the audio into groups of samples and embedding one bit in each of the segments.

This leads to the following situation: If two different bit vectors are embedded in two copies of the same cover with the same key, the two copies differ exactly in those segments where different bits have been embedded as information. Figure 2 shows two embedded bit vectors "01011" and "00001". Both have been embedded in a copy of C. If A and B compare their copies, they find equal segments at position 1,3 and 5 and different segments at position 2 and 4. This enables the attacks we describe in section 3.

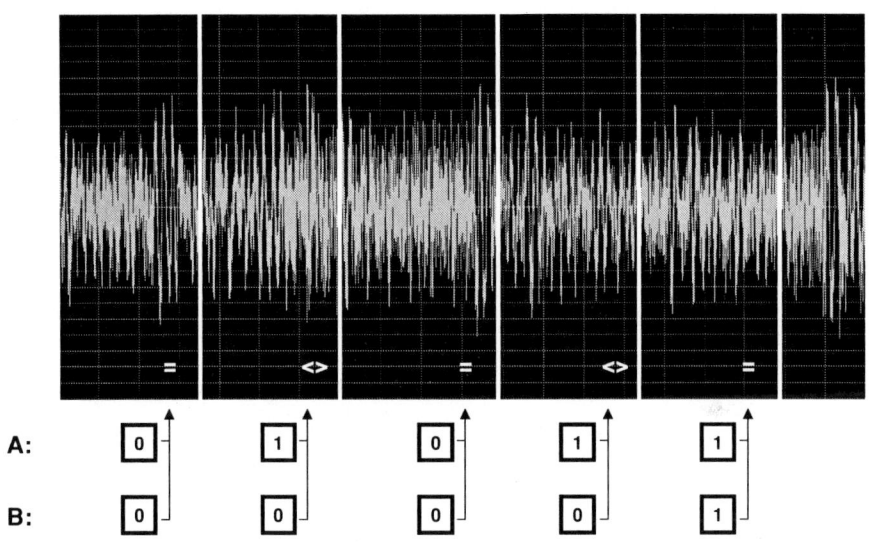

Figure 2: Different embedded bit vectors lead to diffent segments in the copies

3.   SECURITY EVALUATION

In this section we describe our evaluation method for security against combined attacks. First we describe a typical attack scenario where two or more clients work together to remove an embedded watermark. Then we introduce our implemented test tool and its features. We show how this tool is used for simulating attacks and define different evaluation situations.

Only coalition attacks on fingerprint sequences are the subject of our tests in this work. Additionally, all attacks capable of removing the complete watermarking information (the client ID in this case) can be seen as successful attacks. In comparison to the attacks described here, these attacks are oriented on the watermarking algorithm, not against the fingerprinting scheme. For those interested in basic audio watermarking robustness, the Stirmark Benchmark Audio Suite [SMBM] and the corresponding papers [PSR+2001], [SPR+2001] and [SLD02] can be a valuable source of information.

3.1 Attack scenario

For coalition attacks against audio fingerprinting watermarks, we can expect that two or more customers will work together to identify differences between their individual copies received from a common source. One can assume they know that a) the media is protected by a customer identification mechanism and b) to identify customers, differences between the individual copies can be detected. Now they can design attacks based on this knowledge. They will either use an audio editor or a tool to compare their copies and create a new version of the media file where the customer Ids have be obscured.

3.2 Test tool

We have implemented a test tool for coalition attacks which can run several types of attacks on a group of marked copies. The number of individual copies which will be used during the attack can be up to 5, which we think is sufficient for most scenarios.

The basic idea of the tool is to stream the marked copies in parallel and detect differences between the samples. If such differences are detected, out of varying attack strategies we have designed the following difference attacks:

- **Middle**: At detected differences the sample values are added and divided by the number of copies resulting in a middle sample value of all contributing files.
- **Switch**: Out of the detected differences one of the samples is selected. The choice is based on a loop running through the available source files. This creates a mixed sequence of the marked sources.
- **Noise**: The difference positions between the samples are calculated. The difference values at these positions is used as a maximum value for a random change in the first source file. This adds a kind of difference-triggered noise to the source file. A parameter is used to control the amount of added noise.

Table 1: Example results of the different attack modi

	Copy#1	Copy#2	Middle	Mix	Noise	Mosaic
1	-9434	-8813	-9124	-9434	-9678	-9434
2	-544	8087	3772	8087	-8964	-544
3	1261	-5728	-2234	1261	4261	1261
4	4140	-6070	-965	-6070	5642	4140
5	9260	-7917	672	9260	10316	9260
6	-10351	776	-4788	776	-20284	776
7	6641	-163	3239	6641	7290	-163
8	-19931	-13019	-16475	-13019	-22872	-13019
9	3429	2791	3110	3429	3569	2791
10	-1931	-2586	-2258	-2586	-1406	-2586
11	10225	14780	12502	10225	8806	10225
12	-9460	11727	1134	11727	-23591	-9460
13	2520	6243	4382	2520	507	2520
14	9136	-11815	-1340	-11815	19148	9136
15	-15726	-16653	-16190	-15726	-15505	-15726
16	8031	-7181	425	-7181	12338	-7181
17	5499	1078	3288	5499	9245	1078
18	-12781	-5344	-9062	-5344	-13314	-5344
19	-3010	-2480	-2745	-3010	-3467	-2480
20	-9290	1936	-3677	1936	-14671	1936

Furthermore we have included an attack, which is not difference-triggered. It is an audio mosaic attack where first a number of samples of the first source, then of the second source and so on are chosen in a loop. The number of consecutive samples from one source is given as a parameter.

Table 1 provides a short example of the different attack types. 20 samples are taken from two copies (copy#1 and copy#2). The column 'Middle' gives the resulting sample value of the middle attack. For the first samples, this is (-9434 + -8813) / 2 = -9124. 'Mix' is an alternating selection from the samples of copy#1 and #2. For the first sample, copy#1 is chosen. The second sample is taken from copy#2. 'Noise' induces changes in the samples of copy#1 controlled by the difference of the sample values between copy#1 and copy#2. 'Mosaic' is basically a 'mix' attack with a bigger step size, the first 5 samples are taken from copy#1, samples 6 to 10 are taken from copy#2 and so on.

3.3   Test procedure

The test procedure is similar for both evaluated fingerprinting schemes:

1. Create n fingerprint sequences depending on the number of customers.
2. Embed each fingerprint sequence in a cover creating n different copies.
3. Attack the fingerprint by using the n sources with the test tool and creating an attacked copy.
4. Detect the watermark in the attacked copy.
5. Identify the customer described by the detected fingerprinting sequences, verify the correctness, verify transparency.

In figure 3 we see a fingerprint attack scheme with two attackers.

While our tool is able to handle up to five differently marked copies, our tests only include coalition attacks of two or three customers. The applied audio watermarking algorithm is our own prototypic implementation. Tests with available demo versions of commercial products did show no different behaviour, so we use only one algorithm for our tests. Four audio test files have been chosen for evaluation: Classical music, pop, rock and speech to provide the typical range of audio material to be protected. While the performance of watermarking algorithm is not independent from the audio material, it has no direct influence to the results of the fingerprinting tests.

4.   TEST RESULTS

In this section we summarize the test results of the procedure described in section 3. We do not consider transparency tests after embedding, attack and detection. While Middle, Mix and Mosaic attacks did not produce audible artefacts in some initial listening tests, Noise becomes audible at high levels. As Noise is also the least effective attack in our tests, one can assume that an attacker will not chose this method to destroy a fingerprint vector.

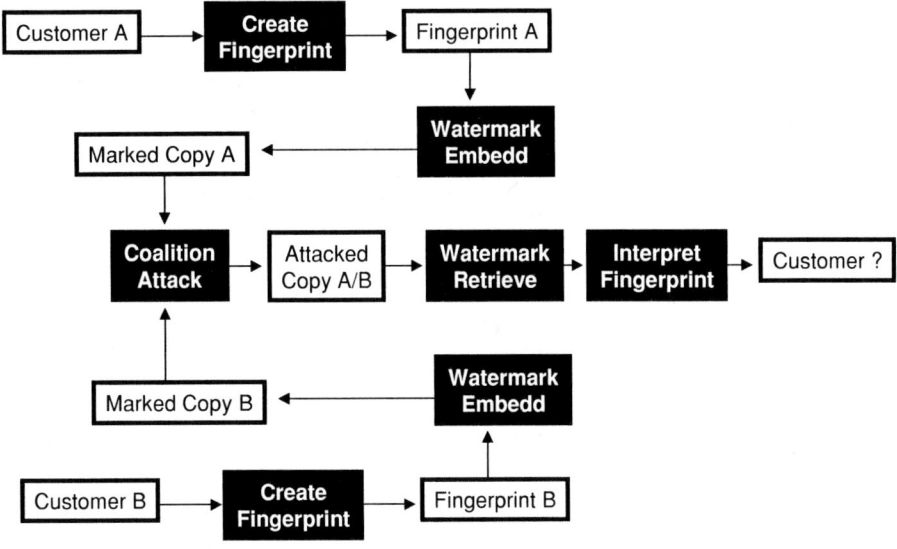

Figure 3: Fingerprint attack scheme with two attackers.

4.1 Schwenk

Table 2 provides an overview of our test results. Line 2 and 3 give the original fingerprint bit vectors, line 4 the result of an AND operation on both bit vectors to identify the common "1" (coalition identifier) bits. We embedded each bit vector twice in one example, the resulting bit vectors for all examples are listed from line 6. From the retrieved fingerprints generated by the Schwenk algorithm we can retrieve the coalition bit (the intersection of bits) successfully to identify the attackers. Most changes in the whole bit vector occur with the middle and switch attacks. The noise attack does not influence the watermark strong enough to result in detecting a wrong bit vector. Therefore if customer A's copy is modulated by the one of customer B, customer A is still identified after the attack. Most problems occur if additional ones are added so that other coalition bits are created. To enable detection, the watermarking algorithm has to ensure that only "0"-values can be created when a "0" and a "1" position are compared.

4.2 Boneh

The Boneh and Shaw algorithm performs similarly to the one by Schwenk when applied with additional bit vector encryption: Middle and Switch attacks result in non-interpretable bit vectors, noise attacks do not affect it in most cases. The most alarming test result is that in some cases a customer not involved in the attack was identified.

Table 2: Test results for client 1 and 2 with a Schwenk et al. fingerprinting scheme

	Bit #		1	2	3	4	5	6	7	8	9	10	11	12	13	14	15	16	17	18	19	20	21	22	23	24	25	26	27	28	29	30	31
	Client	1	0	0	0	0	1	0	0	0	0	0	1	0	0	0	1	0	0	0	1	0	0	1	0	0	0	1	0	0	0	0	0
		2	0	0	0	0	1	0	0	1	0	0	0	0	0	0	0	1	0	0	1	0	0	1	0	0	0	0	0	0	0	1	0
	AND		0	0	0	0	0	0	0	0	0	0	0	0	0	0	0	0	0	0	1	0	0	0	0	0	0	0	0	0	0	0	0
classic	middle	1	0	0	0	0	1	0	1	0	1	0	0	1	0	0	0	1	0	0	1	0	0	1	1	0	0	0	1	0	0	0	0
	middle	2	0	0	0	0	1	0	0	0	1	0	0	1	0	0	0	0	0	0	1	0	0	1	1	0	0	0	0	0	0	1	0
	switch	1	0	0	0	0	1	0	1	0	1	0	0	1	0	0	0	1	0	0	1	0	0	1	1	0	0	0	1	0	0	0	0
	switch	2	0	0	0	0	1	0	0	0	0	0	0	1	0	0	0	0	0	0	1	0	0	1	1	0	0	0	0	0	0	1	0
	noise	1	0	0	0	0	1	0	0	0	0	0	1	0	0	0	1	0	0	0	1	0	0	0	1	0	0	0	1	0	0	0	0
	noise	2	0	0	0	0	1	0	0	0	0	0	1	0	0	0	1	0	0	0	1	0	0	0	1	0	0	0	1	0	0	0	0
pop	middle	1	0	0	0	0	1	1	1	0	0	0	0	0	0	0	1	0	0	0	1	0	0	0	1	0	0	0	0	0	0	0	0
	middle	2	0	0	0	0	1	0	0	1	0	0	0	0	0	0	1	0	0	1	0	0	0	1	0	0	0	0	0	0	0	0	0
	switch	1	0	0	0	0	1	1	1	0	0	0	0	0	0	0	1	0	0	0	1	0	0	0	1	0	0	0	0	0	0	0	0
	switch	2	0	0	0	0	1	0	1	0	1	0	0	0	0	0	1	1	0	0	1	0	0	0	1	0	0	0	1	0	0	0	0
	noise	1	0	0	0	0	1	0	0	0	0	0	1	0	0	0	1	0	0	0	1	0	0	0	1	0	0	0	1	0	0	0	0
	noise	2	0	0	0	0	1	0	0	0	0	0	1	0	0	0	1	0	0	0	1	0	0	0	1	0	0	0	1	0	0	0	0
rock	middle	1	0	0	0	0	1	0	0	0	1	0	0	1	0	0	0	0	0	0	1	0	0	0	0	0	0	0	1	0	0	1	0
	middle	2	0	0	0	0	1	0	0	1	0	0	0	0	0	0	0	0	0	0	1	0	0	0	0	0	0	0	1	0	0	1	0
	switch	1	0	0	0	0	0	0	0	1	0	0	1	0	0	0	0	0	0	0	1	0	0	0	0	0	0	0	1	0	0	1	0
	switch	2	0	0	0	0	1	0	0	1	0	0	0	0	0	0	0	0	0	0	1	0	0	0	0	0	0	0	1	0	0	1	0
	noise	1	0	0	0	0	1	0	0	0	0	0	1	0	0	0	1	0	0	0	1	0	0	0	1	0	0	0	1	0	0	0	0
	noise	2	0	0	0	0	1	0	0	0	0	0	1	0	0	0	1	0	0	0	1	0	0	0	1	0	0	0	1	0	0	0	0
speech	middle	1	0	0	0	0	1	0	0	0	1	0	0	1	0	0	0	1	0	0	1	0	0	1	0	0	0	1	0	0	1	0	0
	middle	2	0	0	0	0	1	0	0	0	0	0	0	0	0	0	1	1	0	0	1	0	0	1	0	0	0	1	0	0	1	0	0
	switch	1	0	0	0	0	1	0	0	0	1	0	0	1	0	0	0	1	0	0	1	0	0	1	0	0	0	1	0	0	1	0	0
	switch	2	0	0	0	0	1	0	0	0	0	0	0	0	0	0	1	1	0	0	1	0	0	1	0	0	0	1	0	0	1	0	0
	noise	1	0	0	0	0	1	0	0	0	0	0	1	0	0	0	1	0	0	0	1	0	0	1	0	0	0	1	0	0	0	0	0
	noise	2	0	0	0	0	1	0	0	0	0	0	1	0	0	0	1	0	0	0	1	0	0	1	0	0	0	1	0	0	0	0	0

Table 3 shows an example result without bit vector encryption. Here we could identify one correct attacker after the attacks. Lines 1 and 2 show the bit vectors after their generation, line 3 to 5 the resulting vectors after the attack. The last column provides the identified customer.

Table 3: Boneh test results without encryption

Process	Bit Vector	Identified
ID 2:	00000000000000000000011111111111111111111111111111111111111111111111111111111111	2
ID 4:	00000000000000000000000000000000000000000000000000000000000001111111111111111111	4
switch	000000000000000000000011010100001011010010111100001011000000011111111111111111111111	4
middle	000000000000000000000011010100010110100101111000010110000001111111111111111111111111	4
noise	00000000000000000000011111111111111111111111111111111111111111111111111111111111	2

We also simulated an attack of three clients against the Boneh fingerprint. Table 4 shows an excerpt of the results. The complete bit vector has a length of 117 bits. In no case one of the clients could be identified. An interesting difference to the attacks with two clients can be observed in bit column 23: Bit #23 has the value 0 in all three client vectors. The detected post-attack bit value became 1 in some cases. These changes occurred several times in the vector, and both from value 1 to 0 and from 0 to 1.

This leads to an additional challenge in client identification compared to two-client attacks as there equal bit position have not been changed. Redundancy and error correction can help to lessen this thread, but also leads to even higher payload requirements.

Table 4: Attack of three clients vs. Boneh

		1	2	3	4	5	6	7	8	9	10	11	12	13	14	15	16	17	18	19	20	21	22	23	24	25	26	27	28	29	30
Client	1	0	1	1	1	1	1	0	1	0	0	0	0	0	1	1	1	1	1	1	1	1	1	1	0	1	0	1	1	1	1
Client	2	0	0	1	0	1	0	0	0	0	0	0	0	0	1	0	1	1	0	1	0	0	0	0	0	1	1	1	0	0	
Client	3	0	0	0	0	1	0	0	0	0	0	0	0	0	0	0	1	1	0	0	0	0	0	0	0	0	0	1	0	0	
AND		0	0	0	0	1	0	0	0	0	0	0	0	0	0	0	1	1	0	0	0	0	0	0	0	0	0	1	0	0	
Middle	1	0	0	1	0	1	0	0	0	0	0	0	0	0	1	0	1	1	0	1	0	0	1	1	0	0	1	1	0	0	
	2	0	0	1	0	1	0	1	0	0	0	0	0	0	0	1	1	1	1	0	0	0	0	0	1	1	1	1			
	3	0	0	1	0	1	0	0	0	0	0	0	0	1	0	1	1	0	1	0	0	0	1	0	1	1	1	0	0		
	4	0	0	1	0	1	0	0	0	0	0	0	0	1	0	1	1	0	1	0	1	0	1	0	0	0	1	1	1	0	0
Mix	1	0	0	1	0	1	0	0	0	0	0	0	0	0	1	0	1	1	0	1	0	0	1	1	0	0	1	1	0	0	
	2	0	0	1	0	1	0	1	0	0	0	0	0	0	0	1	1	1	1	0	0	0	0	0	1	1	1	1			
	3	0	0	1	0	1	0	0	0	0	0	0	0	1	0	1	1	0	1	0	0	0	1	0	1	1	1	0	0		
	4	0	0	1	0	1	0	0	0	0	0	0	0	1	0	1	1	0	1	0	1	0	1	0	0	0	1	1	1	0	0
Noise	1	0	1	1	1	1	1	0	1	0	0	0	0	0	1	1	1	1	1	1	1	1	1	1	0	1	0	1	1	1	1
	2	0	1	1	0	1	0	1	0	1	0	0	0	0	1	1	1	1	1	1	1	1	1	1	0	1	0	1	1	1	1
	3	0	1	1	1	1	1	0	1	0	0	0	0	0	1	1	1	1	1	1	1	1	1	1	0	1	0	1	1	1	1
	4	0	1	1	1	1	1	0	1	0	0	0	0	0	1	1	1	1	1	1	1	1	1	1	0	1	0	1	1	1	1

4.3 Client identification without fingerprinting

To verify the necessity of fingerprinting methods, we also examined the performance of different alternative client identification methods. Three different strategies have been evaluated:

- Same key, different watermark: A customer ID is embedded without a fingerprinting strategy using always the same key. The customer ID could be a real name or an ID number.
- Same watermark, different key: Instead of using the watermark as the carrier, the customer could also be identified by a key. This key depends on the client ID, and if a detector is able to retrieve a watermark from the cover, the client is identified. Searching a customer would require brute force scanning through the key space.
- Different key, different watermark: A combination of both strategies. A key is generated based on the client ID, and the watermarking message is the same client ID. This adds security regarding brute force attacks of third parties as the relation of key and ID may be secret.

We used the same attack tool to evaluate the performance of the three methods and got similar results for all of them. This time 10 test files including the 4 of tests 4.1 und 4.1 have been chosen. Two attackers were simulated. In all cases, a correct detection of the client was possible in less then 50% of the attacks. As in tests 4.1 and 4.2 the results were best with the noise attack. In some attacks and files a correct identification was possible in less then 20% of the cases.

This shows that a strategy not using fingerprinting concepts is very vulnerable to coalition attacks. While the results of both fingerprinting and non-fingerprinting approaches are not satisfying, optimisation potential for fingerprinting-based solutions is much greater. In section 5 we describe one possible optimisation.

5. CONCLUSION AND FUTURE WORK

In comparison to the collusion resistant results of the image and video fingerprinting watermarking algorithms introduced in [DBS+99] and [DHV+01] the audio results show that the coalition bits of the Schwenk et al. approach can be found successfully. Problems occur if "1"s are created by an attack and can cause problems with other possible coalition bits with other customers. Further tests are necessary for a more comprehensive study. Furthermore the Boneh et al. approach has problems with addional ones and it is possible that innocent customer can be identified.

For identifying users that took part in a coalition attack, it could be helpful to change the embedding algorithm so that a rule could be set for mixing two fingerprints. When every time an embedded "0" and "1" are mixed, one specific bit occurs, we would receive a bit vector much more easy to interpret. In the case of the Schwenk algorithm mixing a "0" and a "1" should always result in a "0" as the "1"s are used to identify the group of attackers. E.g. the sequences "0010101" and "0110001" could now result in "0110101" or "0011101" among other possibilities. After an optimisation the only possible result should be "0010001" identifying both attackers by the shared "1"s at bit#2 and bit#6.

To use different embedding strengths for both bits can be a solution and should be take into considerations of further evaluations. In the case of middle or mix attacks this would result in the bit embedded with more strength surviving the coalition attacks. Figure 4 illustrates this concept.

Furthermore generally for a high number of customers the length of the fingerprint vectors for both fingerprint schemes is very high. The optimization of the fingerprint algorithms is an important point for the future research. The problem is to embed the customer vector in material with restricted size.

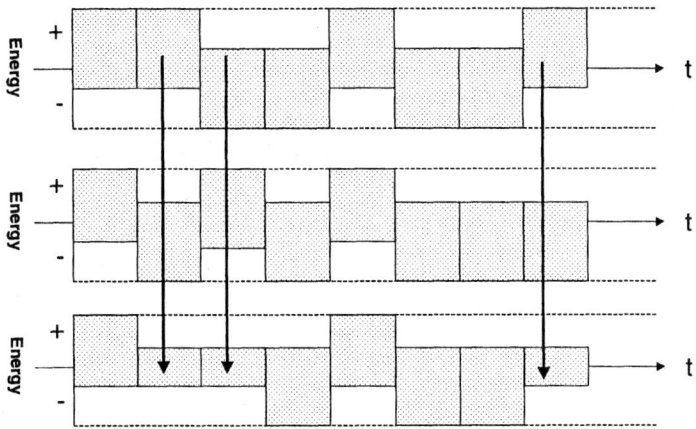

In a common statistical watermarking algorithm, 0 and 1 Bits are embedded by increasing and decreasing energy of frequency bands at the same amount. When the watermarks of line 1 and 2 are combined, the resulting sequence of energy changes becomes hard to interpret.

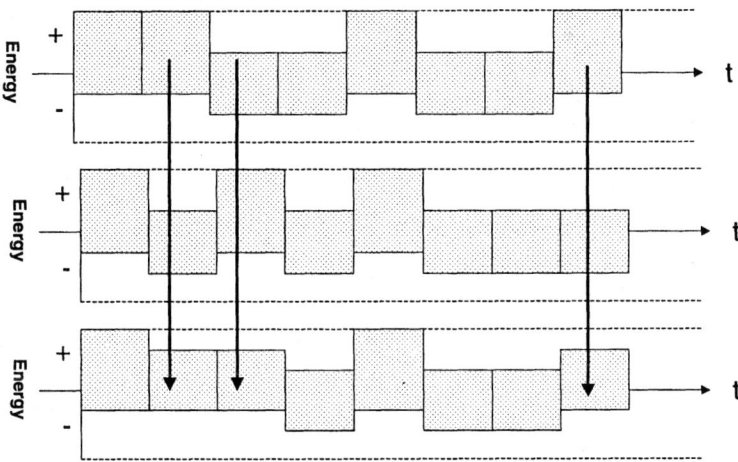

We suggest a stronger increase in energy in one direction, + in this case. Now if line 1 and 2 are combined, the resulting watermark still shows a certain tendency to "+", which is weaker then a original "+" position.

Figure 4: Fingerprinting-optimised watermarking

REFERENCES

[DBS+99]   Dittmann, Jana; Behr, Alexander; Stabenau, Mark; Schmitt, Peter; Schwenk, Joerg; Ueberberg, Johannes (1999), *Combining digital Watermarks and collusion secure Fingerprints for digital Images*, Proceedings of SPIE Vol. 3657, [3657-51], Electronic Imaging '99, San Jose USA, 24-29 January 1999.

[BoSh95]   D. Boneh and J. Shaw, *Collusion-Secure Fingerprinting for Digital Data.* Proc. CRYPTO'95, Springer LNCS 963, S. 452-465, 1995

[BeRo98]   A. Beutelspacher and U. Rosenbaum, *Projective Geometry.* Cambridge University Press 1998.

[DEV+01]   Dittmann, Jana; Hauer, Enrico; Vielhauer, Claus; Schwenk, Jörg; Saar, Eva: Customer Identification for MPEG Video based on Digital Fingerprints. In: Proceedings of Advances in Multimedia Information Processing - PCM 2001, The Second IEEE Pacific Rim Conference on Multimedia, Beijing, China, Springer Verlag, Berlin, pp. 383 - 390, ISBN 3-540-42680-9, 2001.

[Hir98]   J.W.P. Hirschfeld, *Projective Geometries over Finite Fields.* Oxford University Press, 2nd Edition 1998.

[DKL+02]   Jana Dittmann, Stephan Klink, Andreas Lang, Martin Steinebach: Wasserzeichen unterstützte Firewalls, to appear in Proc. Of Enterprise Security, Paderborn, Germany, 2002

[PSR+2001] Petitcolas, F. A. P.; Steinebach, Martin; Raynal, F.; Dittmann, Jana; Fontaine, C.; Fates, N. (2001). Public automated web-based evaluation service for watermarking schemes: StirMark Benchmark. In: Security and Watermarking of Multimedia Contents III, Ping Wah Wong, Edward J. Delp III, Editors, Proceedings of SPIE Vol. 4314, pp. 575 - 584, ISBN 0-8194-3992-4, 2001.

[SPR+2001] Steinebach, Martin; Petitcolas, Fabien A. P.; Raynal, Frederic; Dittmann, Jana; Fontaine, Caroline; Seibel, Christian; Fates, Nazim; Croce Ferri, Lucilla (2001). StirMark Benchmark: Audio watermarking attacks. In: Int. Conference on Information Technology: Coding and Computing (ITCC 2001), April 2 - 4, Las Vegas, Nevada, pp. 49 - 54, ISBN 0-7695-1062-0, 2001.

[SLD02]   Steinebach, Martin; Lang, Andreas; Dittmann, Jana, "StirMark Benchmark: Audio watermarking attacks based on lossy compression", Photonics West 2002, 19 - 25 January 2002, Electronic Imaging 2002: Science and Technology; Multimedia Processing and Applications, Security and Watermarking of Multimedia Contents IV, San Jose, CA, USA

[SMBM]   http://ms-smb.ipsi.thg.de/stirmark/Index.php

SELECTIVE ENCRYPTION OF VISUAL DATA

Classification of Application Scenarios and Comparison of Techniques for Lossless Environments

Champskud J. Skrepth[*]
Carinthia Tech Institute, Austria
9956@edu.fh-kaernten.ac.at

Andreas Uhl[†]
Department of Scientific Computing, Salzburg University
and Carinthia Tech Institute, Austria
uhl@cosy.sbg.ac.at

Abstract   We discuss techniques for selective encryption of visual data. After introducing a classification of application scenarios for selective encryption we derive conditions for the sensible employment of such an approach in each scenario. Finally, we propose and evaluate experimentally several selective encryption approaches for a specific lossless application scenario.

Keywords: Selective image encryption, partial or soft encryption

INTRODUCTION

In the area of multimedia security, the term "soft encryption" is sometimes used as opposed to classical "hard" encryption schemes like AES. Such schemes do not strive for maximum security and trade off security for computational complexity. They are designed to protect multimedia content and fulfill the security requirements for a particular multimedia

---

[*]This artificial name represents a group of students working on this project in the framework of the multimedia 1 laboratory (winterterm 2001/2002).
[†]This work has been partially supported by the Austrian Science Fund, project no. 15170

application. For example, real-time encryption for an entire video stream using classical ciphers requires much computation time due to the large amounts of data involved, on the other hand many multimedia applications require security on a much lower level (e.g. TV broadcasting [6]) or should protect their data just for a short period of time (e.g. news broadcast). Therefore, the search for fast encryption procedures specifically tailored to the target environment is mandatory for multimedia security applications.

Selective or partial encryption (SE) of visual data is an example for such an approach. Here, application specific data structures are exploited to create more efficient encryption systems (see e.g. SE of MPEG video streams [1, 5, 10, 11, 12, 15], of wavelet-based encoded imagery [2, 4, 8, 9, 13, 15], and of quadtree decomposed images [2]). Consequently, SE only protects (i.e. encrypts) the visually most important parts of an image or video representation relying on a secure but slow "classical" cipher.

In this work we discuss selective image encryption techniques for lossless environments. Section 1 provides a classification and discussion of four different application scenarios for SE. We derive conditions for the sensible employment of SE in each scenario and give concrete sample applications. In section 2 we present several different SE approaches for a specific lossless application scenario which are experimentally compared in section 3. In the conclusion we summarize the main results and give an outlook to further work in this direction.

1.   APPLICATION SCENARIOS FOR SELECTIVE IMAGE AND VIDEO ENCRYPTION

Intuitively, SE seems to be a good idea in any case since it is always desirable to reduce the computational demand involved in signal processing applications. However, the security of such schemes is always lower as compared to full encryption. The only reason to accept this drawback are *significant* savings in terms of processing time or power. Therefore, the environment in which SE should be applied needs to be investigated thoroughly in order to decide whether its use is sensible or not.

In the following we discuss a classification of application scenarios for SE of images and videos. The first classification criterion is whether the application operates in a lossless or lossy environment. There exist several reasons why a lossy representation may not be acceptable or necessary:

- Due to requirements of the application a loss of image data is not acceptable (e.g., in medical applications because of reasons related to legal aspects and diagnosis accuracy [14]).

- Due to the low processing power of the involved hardware encoding or decoding of visual data is not possible (e.g. mobile clients).

- Due to the high bandwidth available at the communication channel lossy compression is not necessary.

Note that no matter if lossy or lossless, compression has always to be performed prior to encryption since the statistical properties of encrypted data prevent compression from being applied successfully. Moreover, the reduced amount of data after compression decreases the computational demand of the subsequent encryption stage.

The second classification criterion is whether the image data is given as plain image data or in form of a bitstream resulting from prior compression. For example, in on-line applications the plain image data may be accessed directly after being captured by a digitizer before being compressed. On the other hand, as soon as visual data has been stored or transmitted it has been compressed in some way which is true for most off-line applications. The following table introduces four types of possible SE application scenarios based on the criteria introduced so far which will be discussed subsequently.

Table 1. Application scenarios for selective encryption

	Lossy	Lossless
Bitstream	Scenario A	Scenario B
Image	Scenario C	Scenario D

In the following, t denotes the time required to perform an operation, E is the encryption function, SE the selective encryption function, C is compression, P is the preprocessing involved in the selective encryption scheme (where P means the extraction of relevant features), and $>>$ means significantly larger. Please note that the processing time t is not equivalent to computational complexity: for example, if compression is performed in hardware and encryption in software, the time required for compression will be considerably lower as for encryption, contrasting to the relation if both operations are performed in software (compare scenario C).

- **Scenarios A and B:** Given the bitstream B resulting from prior compression, the following condition must be fulfilled in order to justify the use of SE:

$$t(E(B)) >> t(P) + t(SE(B)) \tag{1}$$

In this case, P is the identification of relevant features in the compressed bitstream. Depending on the type of bitstream, $t(P)$ may range from negligibly small to a considerably large amount of time. If the bitstream is embedded or is composed of several quality layers, the identification of parts subjected to SE is straightforward ($t(P) = 0$) – the first part of the embedded bitstream or the base layer is encrypted only. In case of a less structured bitstream it might be necessary to partially decode or at least parse the bitstream to identify significant features (e.g. DC or large AC coefficients of a JPEG encoded image), thereby causing $t(P)$ to increase linearly with $t(SE(B))$. However, given an embedded bitstream the condition $t(E(B)) >> t(SE(B))$ and therefore equation (1) may be satisfied easily and SE definitely makes sense. Concrete applications are video on demand for scenario A and retrieval of medical images from a database for scenario B.

- **Scenario C:** Given the raw image data I, the following condition must be fulfilled in order to justify the use of SE:

$$t(C(I)) + t(E(C(I))) >> t(C(I)) + t(P) + t(SE(C(I))) \tag{2}$$

As in scenarios A and B, P is again the identification of relevant features in the compressed bitstream and the same considerations about its execution time apply here. In any case, even if $t(P) = 0$, equation (2) is very hard to satisfy since $t(C(I)) >> t(E(C(I)))$ holds for most lossy coding schemes and symmetrical ciphers if both schemes are executed in software. Therefore, the difference between $t(E(C(I)))$ and $t(SE(C(I)))$ often does not matter in practice. This effect is even more pronounced for high compression ratios of course (since the resulting bitstream after compression is already rather small). Consequently, given the raw image data, the decrease in terms of security often does not justify the marginal savings in processing time as achieved by SE in a software based system. Concrete applications for this scenario are videoconferencing and on-line surveillance.

- **Scenario D:** Given the raw image data I, the following condition must be fulfilled in order to justify the use of SE:

$$t(E(I)) >> t(P) + t(SE(I)) \tag{3}$$

In contrast to scenarios A - C, encryption is applied directly to the raw image data. Note that if $t(C(I)) + t(E(C(I))) > t(E(I))$ does not hold, the image data is compressed using a lossless codec and the considerations of scenario B apply. Usually, this is not the case since $t(C(I)) >> t(E(I))$ is valid also for almost all lossless codecs and symmetrical ciphers. Additionally, the data reduction of lossless schemes is much lower as compared to lossy ones making the contribution of $t(E(C(I)))$ significant as well. When applying encryption to the raw image data, P denotes the identification of relevant features in the raw image data which may be done in various ways (see next section for some examples). However, it is crucial that $t(P)$ is not too large to satisfy equation (3). If $t(P)$ can be made small, SE is a sensible approach in this setting. A concrete sample application for this scenario is teleradiology with mobile clients to enable fast and exact on-site diagnosis.

Most SE schemes discussed in literature involve lossy coding schemes and are therefore categorized into scenarios A or C. Regarding the analysis given above, it makes a big difference with respect to the usefulness of SE whether the raw image data or a compressed bitstream is given – this fact is usually ignored which is an obvious shortcoming in many papers on SE. Consequently, it is important to analyse the concrete application setting in which SE should be used in order to judge its appropriateness. In the following section we discuss several approaches to extract significant image features (i.e. preprocessing approaches) for SE schemes in application scenario D.

2. SELECTIVE IMAGE ENCRYPTION TECHNIQUES IN LOSSLESS ENVIRONMENTS

SE applied in application scenario D consists of two stages: preprocessing P which extracts significant features of the image and the selective encryption process $SE(I)$ itself which encrypts those features. Therefore, two components of the overall execution time of such a scheme need to be investigated: $t(P)$ and $t(SE(I))$ where the latter is directly proportional to the percentage of data encrypted in this stage (perc) relative to the original data amount. Following the IEEE standard data types we assume the image to be given in 8bit/pixel (bpp) precision (char), whereas int and float data are assumed to require 32 bpp.

2.1. Spatial domain techniques

2.1.1 Multiresolution pyramids. As a first step we construct a quater-sized version of the image ("approximation") using a 4-pixel average (AV), a 4-pixel median (ME), or subsampling by 2 in each direction (DS). Subsequently, we construct a full-sized version of the image ("prediction") by using an interpolated version of the approximation only (different types of interpolation are used: linear (default setting) and cubic (CI)). This prediction is subtracted from the original image resulting in the "residual". The approximation is encrypted and transmitted together with the residual in plaintext. The construction of the approximation is iterated to construct smaller versions. Whereas the bitdepth is not influenced by these operations (the residual can usually be represented as `signed char` requiring 8bpp as well) the overall amount of data to be transmitted is. After one iteration we result in 125% of the original data (residual plus quater-sized approximation), 106.25% after the second iteration, etc. There is only a small amount of sensible values attained for `perc`: 25, 6.25, 1.56, ..., $t(P) \neq 0$ of course but it is reasonably small using our simple approach.

2.1.2 Bitplane encryption. We consider the 8bpp data in the form of 8 bitplanes, each bitplane associated with a position in the binary representation of the pixels. The SE approach is to encrypt a subset of the bitplanes only, starting with the bitplane containing the MSB of the pixels. Each possible subset of bitplanes may be chosen for SE, however, the minimal value for `perc` is 12.5 (when encrypting the MSB bitplane only) increasing in steps of 12.5. $t(P)$ is negligible using this approach which is denoted BP1. Alternatively, the Gray-code binary representation can be used (BP2) which causes $t(P) \neq 0$ due to the conversion operation which has a complexity order of magnitude similar to an addition operation for each pixel. The encrypted bitplanes are transmitted together with the remaining bitplanes in plain text.

2.2. Transform domain techniques

In contrast to spatial domain methods the operation P (i.e. the transform) increases the bitdepth significantly. No matter if `float` or `int` data are used in the transform, we result in 400% of the original data after the transform. Consequently, if we encrypt 25% of the coefficients for example, `perc` is 100. Additionally, $t(P) \neq 0$ and it is considerably large. Note that for all techniques discussed the amount of data to be transmitted is 400% of the original image data.

2.2.1 DCT. The DCT is well known to extract global image characteristics efficiently and is used for watermarking applications for these reasons (see e.g. Cox's scheme [3]). We use the DCT in two flavours: as full frame DCT (DCT1) and as DCT applied to 8 × 8 pixels blocks (DCT2) due to complexity reasons. Following the zig-zag scan (compare e.g. JPEG) we encrypt the first coefficients or the first coefficients from each block, respectively. The encrypted coefficients are transmitted together with the non-encrypted ones. Given a 512 × 512 pixels image and using DCT2, the lowest value for `perc` is 6.25 (i.e. the DC coefficient is encrypted only for each block) and increasing in steps of 6.25 per additional coefficient, whereas `perc` may be set almost arbitrarily with DCT1.

2.2.2 Wavelet transform. In many applications wavelet transforms (WT) compete with and even replace the DCT due to their improved localization properties (e.g., the WT is used in many watermarking schemes [7]). We use the Haar transform due to complexity reasons and investigate two different SE approaches. For both schemes the decomposition depth is a parameter, WT1 subsequently encrypts the approximation subband only, whereas WT2 encrypts both approximation subband and a number of additional significant coefficients (i.e. larger than a threshold) from other subbands. WT1 delivers only a small amount of sensible values for `perc` (25, 6.25, ...), whereas `perc` may be set almost arbitrarily with WT2.

3. EXPERIMENTS

The aim of the experimental section is twofold. First, we want to evaluate the effectiveness of the proposed SE schemes with respect to execution efficiency. Second, and most important, the security of these schemes is assessed and compared. Assuming the cipher in use is unbreakable we conduct a simple ciphertext-only attack by reconstructing the selectively encrypted images. The encrypted parts would introduce noise-type distortions in directly reconstructed images. Therefore, we replace the encrypted parts by artificial data mimicing typical images (see section 3.1 for details). Subsequently, reconstruction is performed as usual, treating the encrypted and replaced parts as being non-encrypted. A major shortcoming of many SE investigations is the lack of quantifying the quality of the visual data that can be obtained by attacks against SE. Mostly visual examples are provided only. Here, the quality of the obtained images is assessed using PSNR. Additionally, we relate the numerical values to visual examples.

3.1.   Experimental settings

We use the classical 8bpp, 512×512 pixels Lena grayscale image as testimage. All SE schemes are implemented using MATLAB®, as cipher we use an AES implementation specifically developed for this particular purpose with blocksize 128 bit and a 128 bit key. The 128 bit block of AES is filled with data according to data types: a quater of a line when encrypting bitplanes, 16 consecutive pixels of a line when encrypting char and 4 values when encrypting float transform coefficients.

In order to conduct our ciphertext-only attacks the encrypted data needs to be replaced for reconstruction as explained in the previous section. In the case of multiresolution pyramids, the encrypted approximation is replaced by an approximation consisting of constant grayvalue 128. Similarly, the encrypted wavelet approximation subband is replaced by an approximation subband resulting from decomposing an image with constant grayvalue 128. The encrypted DCT coefficients are replaced by a linearly decreasing sequence of coefficients, starting from a DC coefficient again obtained from transforming an image with constant grayvalue 128, terminating at the first non-encrypted original coefficient. Bitplane encryption is attacked by replacing the encrypted bitplane with a constant 0 bitplane and additionally compensating for the decrease in average luminance by adding 64 to each pixel if only the MSB bitplane was encrypted, 96 if the MSB and next bitplane have been encrypted, and so on. The efficiency of these simple replacements is shown in Figure 4. The images 4.a and 4.c are directly reconstructed, whereas 4.b and 4.d are reconstructed using the replacement strategy.

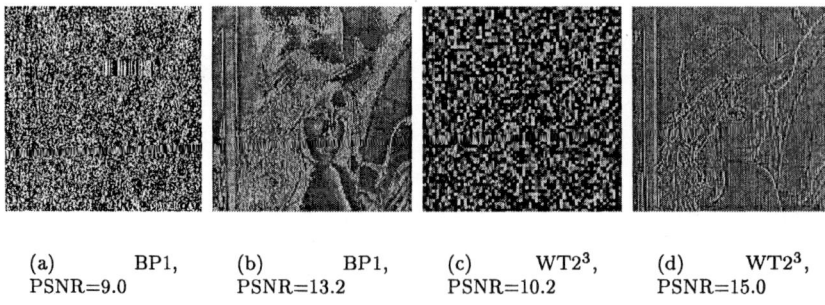

(a)   BP1,   (b)   BP1,   (c)   WT2³,   (d)   WT2³,
PSNR=9.0   PSNR=13.2   PSNR=10.2   PSNR=15.0

Figure 1.   Visual examples for the efficiency of the replacement operation, perc=25.

Obviously, not only the visual appearance but also the numerical PSNR values have been significantly improved by the replacement strategy. These effects are equivalently observed for multiresolution pyramid and DCT based methods.

3.2. Experimental results

Equation 4 relates the time demand of the preprocessing stage $t(P)$ of the various SE schemes introduced in our implementation. For schemes based on multiresolution pyramids or WT, superscripts denote the number of decomposition iterations performed and CI denotes cubic interpolation to generate the prediction in the case of multiresolution pyramids.

$$0 = BP1 < BP2 << DCT2 < WT1^i < WT2^i <$$
$$DS^i < AV^i < ME^i < ME^i(CI) << DCT1 \qquad (4)$$

Note that the computational effort to encrypt the entire image with AES is about two times as high as for performing a wavelet transform and is comparable to the full frame DCT for the image size considered. This fact makes the approach DCT1 useless for practical applications unless the DCT is performed in hardware and AES in software. For all techniques but the bitplane encryption methods, sensible performance gains compared to full encryption can consequently be expected for `perc` ≤ 25 only. Bitplane encryption is more efficient than full AES encryption even for `perc` $= 87.5$ due to its low complexity preprocessing stage P.

$t(SE(I))$ is entirely independent of $t(P)$ and is assessed in the following by directly relating the amount of data encrypted in this stage (expressed in `perc`, see section 2 for a definition) to the PSNR of the reconstructed images. Note that "better" methods have lower PSNR values !

In the case of multiresolution pyramids Figure 2.a shows the extremal curves for DS and ME. All other techniques fall within this range. Cubic interpolation CI enhances the result of simple downsampling DS significantly which is not true for the median approximation ME.

Figure 2.b compares the two flavors of bitplane encryption and DCT, respectively. The Gray-code representation shows better results as compared to the classical binary code. Surprisingly, both approaches do not exhibit monotonically decreasing PSNR values for increasing `perc`, BP1 even has increasing ones. The DCT based methods are clearly inferior to bitplane encryption, DCT1 which is based on global DCT is 2 dB and more inferior to DCT2. Finally we compare the wavelet transform techniques in Figure 3.a. WT1 which simply encrypts the approximation subband is superior to both techniques relying on encrypting approximation data plus significant coefficients.

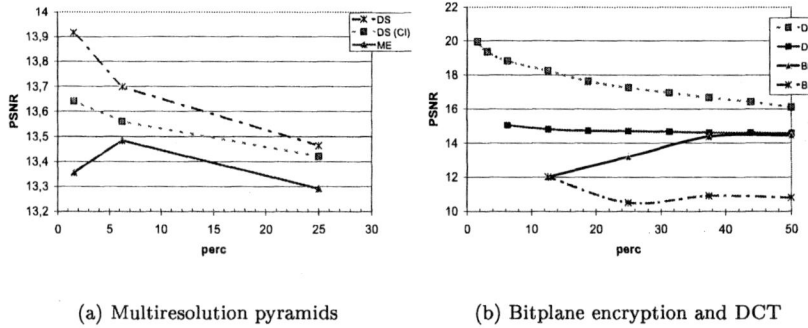

(a) Multiresolution pyramids   (b) Bitplane encryption and DCT

Figure 2.   Recovering selectively encrypted images

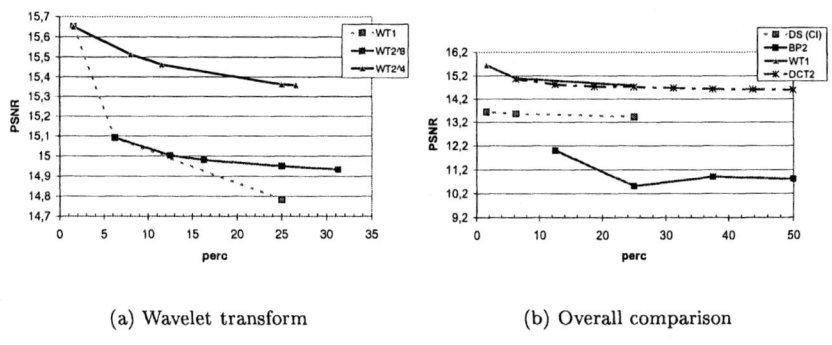

(a) Wavelet transform   (b) Overall comparison

Figure 3.   Recovering selectively encrypted images

For an overall comparison, we have selected the best performing SE algorithm from each group (i.e. that with lowest PSNR), except for the multiresolution pyramid techniques since it has turned out that the visual performance of DS(CI) is significantly better (i.e. worse image quality) as compared to ME (see Figures 4.a and 4.b). In Figure 3.b we notice that the transform based techniques can not compete with the spatial domain techniques. This seems to be surprising at first – note that this is partially due to the data expansion caused by the transforms. However, from the plot we see that this is definitely not the only reason since even compensating the factor 4 expansion does not make WT1 and DCT2 competitive to BP2 and DS(CI). BP2 outperforms DS(CI) with PSNR values up to 2 dB lower. Consequently, in terms of pure PSNR performance, bitplane encryption using Gray-code representation is the best SE technique of the proposed ones. In the following we will relate these findings to visual perception.

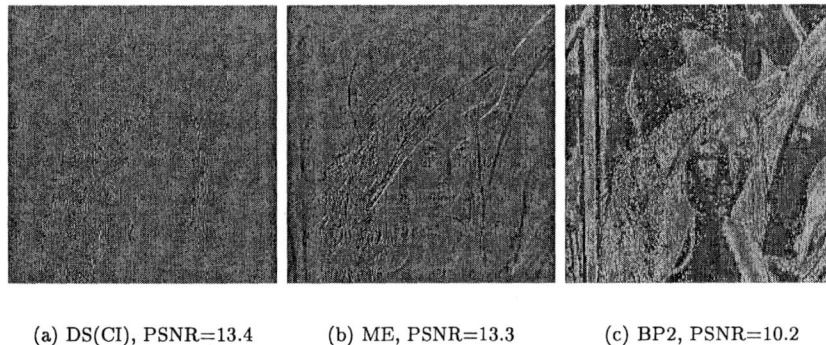

(a) DS(CI), PSNR=13.4   (b) ME, PSNR=13.3   (c) BP2, PSNR=10.2

Figure 4.   Visual examples of recovered images after attack, perc=25.

Figure 4 compares the visual appearance of recovered images which have been subject to spatial domain selective encryption at a rate of perc=25 (i.e. one quater of the original data is encrypted). We face a severe mismatch between PSNR and perceived quality – whereas almost no information is visible at 13.4 dB with DS(CI) (Figure 4.a), we clearly see some edges of the Lena image at 13.3 dB with ME (Figure 4.b). The extremely low PSNR values of BP2 (10.2 dB in this case, Figure 4.c) do not correspond with the visual perception at all. Whereas the luminance appearance is significantly alienated, the objects in the image are perceived clearly.

The visual appearance of recovered images which have been subject to transform domain selective encryption is compared in Figure 5.

The poor performance of DCT1 is visually confirmed, also the similar performance of DCT2 and WT1 in terms of PSNR corresponds well to perception. However, the 1.5dB difference between ME and DCT2/WT1 can not be confirmed visually, all three reconstructions show an almost equal amount of edge information. Note also that the PSNR computed between the image Lena and its entirely AES encrypted version is 9.2 dB whereas PSNR between Lena and an image with constant grayvalue 128 is 14.5 dB ! Both images do not carry any structural information related to Lena, however, the PSNR values differ more than 5 dB. Also, the value 14.5 dB is higher (i.e. more similar to Lena) as compared to the values corresponding to Figures 4.a and 4.b where at least some edge information is visible.

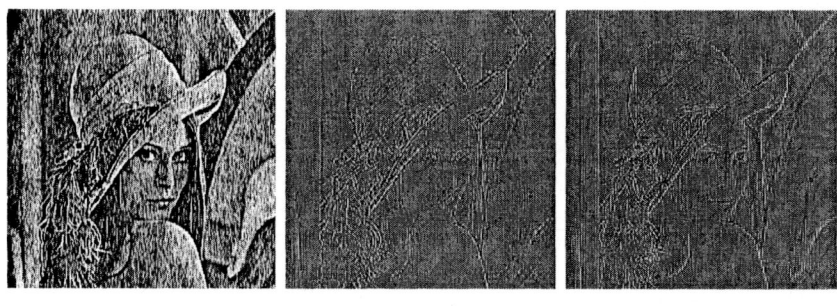

(a) DCT1, PSNR=16.7   (b) DCT2, PSNR=14.8   (c) WT1, PSNR=14.8

Figure 5.   Visual examples of recovered images after attack, perc=25.

Consequently, we may state that PSNR and visual perception do not correspond at all at low quality. Therefore, PSNR is not suited to evaluate SE schemes where the aim is to achieve the lowest image quality possible. Only SE schemes not suited at all for SE (e.g. DCT1) may be identified correctly using PSNR.

When trying to combine both aspects, visual perception and PSNR, in order to identify the best SE technique, we find the multiresolution pyramid technique DS(CI) to be the winner from a pure security point of view. However, this approach is not very scalable since already the second most secure mode (perc=6.25) already reveals a certain amount of edge information. In cases where severe alienation is sufficient to protect the data, bitplane encryption based on Gray-code representation is the method of choice, since this technique is very fast as well.

4.   CONCLUSION

After introducing a classification for selective encryption application scenarios we have investigated several techniques suited for lossless environments with low computing capacity (e.g. mobile clients). From a pure security point of view a specific multiresolution technique in the spatial domain has turned out to be the best method, if alienation is sufficient and speed important, bitplane encryption based on Gray-code representation is a possible choice as well. We have found that PSNR is hardly useful in order to rate SE schemes, consequently empirical methods involving human viewers are required for this aim at present. In future work we will accomplish this to assess the proposed schemes more accurately and we will investigate alternative numerical measures (e.g. picture quality scale) to evaluate low quality visual data consistently. Additionally, we will develop more scalable multiresolution pyramid methods (especially focusing on DS(CI)) by allowing approximations of arbitrary size.

References

[1] A. M. Alattar, G. I. Al-Regib, and S. A. Al-Semari. Improved selective encryption techniques for secure transmission of MPEG video bit-streams. In *Proceedings of the 1999 IEEE International Conference on Image Processing (ICIP'99)*. IEEE Signal Processing Society, 1999.

[2] H. Cheng and X. Li. Partial encryption of compressed images and videos. *IEEE Transactions on Signal Processing*, 48(8):2439–2451, 2000.

[3] Ingemar J. Cox, Joe Kilian, Tom Leighton, and Talal G. Shamoon. Secure spread spectrum watermarking for multimedia. In *Proceedings of the IEEE International Conference on Image Processing, ICIP '97*, volume 6, pages 1673–1687, Santa Barbara, California, USA, October 1997.

[4] Raphaël Grosbois, Pierre Gerbelot, and Touradj Ebrahimi. Authentication and access control in the JPEG 2000 compressed domain. In A.G. Tescher, editor, *Applications of Digital Image Processing XXIV*, volume 4472 of *Proceedings of SPIE*, San Diego, CA, USA, July 2001.

[5] Thomas Kunkelmann. Applying encryption to video communication. In *Proceedings of the Multimedia and Security Workshop at ACM Multimedia '98*, pages 41–47, Bristol, England, September 1998.

[6] Benoit M. Macq and Jean-Jacques Quisquater. Cryptology for digital TV broadcasting. *Proceedings of the IEEE*, 83(6):944–957, June 1995.

[7] P. Meerwald and A. Uhl. A survey of wavelet-domain watermarking algorithms. In Ping Wah Wong and Edward J. Delp, editors, *Proceedings of SPIE, Electronic Imaging, Security and Watermarking of Multimedia Contents III*, volume 4314, San Jose, CA, USA, January 2001. SPIE.

[8] A. Pommer and A. Uhl. Wavelet packet methods for multimedia compression and encryption. In *Proceedings of the 2001 IEEE Pacific Rim Conference on Communications, Computers and Signal Processing*, pages 1–4, Victoria, Canada, August 2001. IEEE Signal Processing Society.

[9] A. Pommer and A. Uhl. Selective encryption of wavelet packet subband structures for obscured transmission of visual data. In *Proceedings of the 3rd IEEE Benelux Signal Processing Symposium (SPS 2002)*, pages 25–28, Leuven, Belgium, March 2002. IEEE Benelux Signal Processing Chapter.

[10] Lintian Qiao and Klara Nahrstedt. Comparison of MPEG encryption algorithms. *International Journal on Computers and Graphics (Special Issue on Data Security in Image Communication and Networks)*, 22(3):437–444, 1998.

[11] C. Shi and B. Bhargava. A fast MPEG video encryption algorithm. In *Proceedings of the ACM Multimedia 1998*, pages 81–88, Boston, USA, 1998.

[12] L. Tang. Methods for encrypting and decrypting MPEG video data efficiently. In *Proceedings of the ACM Multimedia 1996*, pages 219–229, Boston, USA, November 1996.

[13] T. Uehara, R. Safavi-Naini, and P. Ogunbona. Securing wavelet compression with random permutations. In *Proceedings of the 2000 IEEE Pacific Rim Conference on Multimedia*, pages 332–335, Sydney, December 2000. IEEE Signal Processing Society.

[14] S. Wong, L. Zaremba, D. Gooden, and H.K. Huang. Radiologic image compression – a review. *Proceedings of the IEEE*, 83(2):194–219, 1995.

[15] Wenjun Zeng and Shawmin Lei. Efficient frequency domain video scrambling for content access control. In *Proceedings of ACM Multimedia 1999*, pages 285–293, Orlando, FL, USA, November 1999.

BIOMETRIC AUTHENTICATION — SECURITY AND USABILITY

Václav Matyáš and Zdeněk Říha
Faculty of Informatics, Masaryk University Brno, Czech Republic
{matyas, zriha} @fi.muni.cz

Abstract   We would like to outline our opinions about the usability of biometric authentication systems. We outline the position of biometrics in the current field of computer security in the first section of our paper. The second chapter introduces a more systematic view of the process of biometric authentication – a layer model (of the biometric authentication process). The third section discusses the advantages and disadvantages of biometric authentication systems. We also propose a classification of biometric systems that would allow us to compare the biometrics systems reasonably, along similar lines to Common Criteria [1] or FIPS 140-1/2 [4]. We conclude this paper with some suggestions where we would suggest to use biometric systems and where not.

Keywords: authentication, biometrics, classification, evaluation, security.

1.   INTRODUCTION

This paper summarises our opinions and findings after several years of studying biometric authentication systems and their security. Our research on security and reliability issues related to biometric authentication started in 1999 at Ubilab, the Zurich research lab of bank UBS, and has been continuing at the Masaryk University Brno since mid-2000. This paper summarises our personal views and opinions on pros and cons of biometric authentication in computer systems and networks.

Proper *user identification/authentication* is a crucial part of the access control that makes the major building block of any system's security. User identification/authentication has been traditionally based on:

* something that the user *knows* (typically a PIN, a password or a passphrase) or

* something that the user *has* (e.g., a key, a token, a magnetic or smart card, a badge, a passport).

These traditional methods of the user authentication unfortunately do not authenticate the *user* as such. Traditional methods are based on properties that can be forgotten, disclosed, lost or stolen. Passwords often are easily accessible to colleagues and even occasional visitors and users tend to pass their tokens to or share their passwords with their colleagues to make their work easier. Biometrics, on the other hand, authenticate humans as such – in case the biometric system used is working properly and reliably, which is not so easy to achieve. Biometrics are automated methods of identity verification or identification based on the principle of measurable physiological or behavioural characteristics such as a fingerprint, an iris pattern or a voice sample. Biometric characteristics are (or rather should be) unique and not duplicable or transferable. While the advantages of biometric authentication definitely look very attractive, there are also many problems with biometric authentication that one should be aware of.

2. THE LAYER MODEL

Although the use of each biometric technology has its own specific issues, the basic operation of any biometric system is very similar. The separation of actions can lead to identifying critical issues and to improving security of the overall process of biometric authentication. The layer model was designed by our biometrics team (the authors, Hans-Peter Frei, Kan Zhang) during the Ubilab biometrics project, and its structure is also similar to some findings presented in other seminal works on biometric authentication (e.g., [3, 5]).

The whole process starts with the enrolment:

2.1 First measurement (acquisition)

This is the first contact of the user with the biometric system. The user's biometric sample is obtained using an input device. Quality of the first biometric sample is crucial for further authentications of this user. It may happen that even multiple acquisitions do not generate biometric samples with sufficient quality. Such a user cannot be registered with the system. There are also mute people, people without fingers or with injured eyes. Both these categories create a 'fail to enrol' (FTE) group of users. Users very often do not have any previous experience with the kind of the biometric system they are being registered with, so the first measurement should be guided by a professional who explains the use of the biometric reader.

2.2 Creation of master characteristics

The biometric measurements are processed after the acquisition. The number of biometric samples necessary for further processing is based on the nature of given biometric technology. Sometimes a single sample is sufficient, but often multiple (usually 3 or 5) biometric samples are required. The biometric characteristics are most commonly neither compared nor stored in the raw format (say as a bitmap).

2.3 Storage of master characteristics

After processing the first biometric sample(s) and extracting the features, we have to store (and maintain) the newly obtained master template. Choosing proper discriminating characteristic for the categorisation of records in large databases can improve identification (search) tasks later on. There are basically 4 possibilities where to store the template: in a card, in the central database on a server, on a workstation or directly in an authentication terminal. The storage in an authentication terminal cannot be used for large-scale systems, in such a case only the first two possibilities are applicable. If privacy issues need to be considered then the storage on a card (magnetic stripe, smart or 2D bar) has an advantage, because in this case no biometric data must be stored (and potentially misused) in a central database.

As soon as the user is enrolled, she can use the system for successful authentications or identifications. This process is typically fully automated and takes the following steps:

2.4 Acquisition(s)

Current biometric measurements must be obtained for the system to be able to make comparison with the master template. These subsequent acquisitions of the user's biometric measurements are done at various places where authentication of the user is required. It is often up to the reader to check that the measurements obtained really belong to a live persons (the liveness property). In many biometric techniques (e.g., fingerprinting) the further processing trusts the biometric hardware to check the liveness of the person and provide genuine biometric measurements only. Some other systems (like the face recognition) check the user's liveness in software (time-phased sampling).

2.5 Creation of new characteristics

The biometric measurements obtained in the previous step are processed and new characteristics are created. Only a single biometric sample is usually available. This might mean that the number or quality of extracted features is lower than at the time of enrolment.

2.6 Comparison

Currently computed characteristics are compared with the characteristics obtained during enrolment. If the system performs (identity) verification then these newly obtained characteristics are compared only to the master template. For an identification request the new characteristics are matched against a large number of master templates.

2.7 Decision

The final step in the verification process is the yes/no decision based on a threshold. This security threshold is either a parameter of the matching process or the resulting score is compared with the threshold value. Although the error rates quoted by manufactures (typical values of equal error rate (ERR)[1] do not exceed 1%) might indicate that biometric systems are very accurate, the reality is much worse. Especially the false rejection rate is quite high (very often over 10%) in real applications. This prevents legitimate users to gain their access rights and stands for a significant problem of biometric systems.

3. WHAT ARE THE ADVANTAGES OF BIOMETRIC AUTHENTICATION

The primary advantage of biometric authentication methods over other methods of user authentication is that they really do what they should, i.e., they *authenticate the user.* These methods use real human physiological or behavioural characteristics to authenticate users. These biometric characteristics are (more or less) permanent and not changeable. It is also not easy (although in some cases not principally impossible) to change one's fingerprint, iris or other biometric characteristics.

Users cannot pass their biometric characteristics to other users as easily as they do with their cards or passwords.

Biometric objects cannot be stolen as tokens, keys, cards or other objects used for the traditional user authentication, yet biometric characteristics can be stolen from computer systems and networks. Biometric characteristics are not secret and therefore the availability of a user's fingerprint or iris pattern does not break security the same way as availability of the user's password. Even the use of dead or artificial biometric characteristics should not let the attacker in.

Most biometric techniques are based on something that cannot be lost or forgotten. This is an advantage for users as well as for system administrators because the problems and costs associated with lost, reissued or temporarily issued tokens/cards/passwords can be avoided, thus saving some costs of the system management.

Another advantage of biometric authentication systems may be their *speed*. The authentication of a habituated user using an iris-based identification system may take 2 (or 3) seconds while finding your key ring, locating the right key and using it may take some 5 (or 10) seconds.

3.1   Disadvantages of biometric authentication

So why do not we use biometrics everywhere instead of passwords or tokens? Nothing is perfect, and biometric authentication methods also have their own shortcomings. First of all the performance of biometric systems is not ideal (yet?). Biometric systems still need to be improved in the terms of accuracy and speed. Biometric systems with the false rejection rate under 1% (together with a reasonably low false acceptance rate) are still rare today. Although few biometric systems are fast and accurate (in terms of low false acceptance rate) enough to allow identification (automatically recognising the user identity), most of current systems are suitable for the verification only, as the false acceptance rate is too high[2].

The fail to enrol rate brings up another important problem. Not all users can use any given biometric system. People without hands cannot use fingerprint or hand-based systems[3]. Visually impaired people have difficulties using iris or retina based techniques. As not all users are able to use a specific biometric system, the authentication system must be extended to handle users falling into the FTE category. This can make the resulting system more complicated, less secure or more expensive. Even enrolled users can have difficulties using a biometric system. The FTE rate says how many of the input samples are of insufficient quality. Data acquisition must be repeated if the quality of input sample is not sufficient for further processing and this would be annoying for users.

Biometric data are not considered to be secret and security of a biometric system cannot be based on the secrecy of user's biometric characteristics. The server cannot authenticate the user just after receiving her correct biometric characteristics. The user authentication can be successful only when user's characteristics are fresh and have been collected from the user being authenticated. This implies that the biometric input device must be trusted. Its authenticity should be verified (unless the device and the link are physically secure) and user's liveness would be checked. The input device also should be under human supervision or tamper-resistant. The fact that biometric characteristics are not secret brings some issues that traditional authentication systems need not deal with. Many of the current biometric systems are not aware of this fact and therefore the security level they offer is limited.

Some biometric sensors (particularly those having contact with users) also have a *limited lifetime*. While a magnetic card reader may be used for years (or even decades), the optical fingerprint reader (if heavily used) must be regularly cleaned and even then the lifetime need not exceed one year.

Biometric systems may violate user's *privacy*. Biometric characteristics are sensitive data that may contain a lot of personal information. The DNA (being the typical example) contains (among others) the user's preposition to diseases. This may be a very interesting piece of information for an insurance company. The body odour can provide information about user's recent activities. It is also told [3] that people with asymmetric fingerprints are more likely to be homosexually oriented, etc.

Use of biometric systems may also imply loss of anonymity. While one can have multiple identities when authentication methods are based on something the user knows or has, biometric systems can sometimes link all user actions to a single identity.

Biometric systems can potentially be quite troublesome for some users. These users find some biometric systems *intrusive* or personally invasive. Even if no biometric system is really dangerous, users are occasionally afraid of something they do not know much about. In some countries people do not like to touch something that has already been touched many times (e.g., biometric sensor), while in some countries people do not like to be photographed or their faces are completely covered.

Lack of *standards* (or ignorance of standards) may also posses a serious problem. Two similar biometric systems from two different vendors are not likely to interoperate at present.

4. POSSIBLE CLASSIFICATION OF BIOMETRIC SYSTEMS

Classifications help to compare systems. The famous Orange Book [2] divided systems into four categories (A – D) with additional subcategories. All the security features (such as access control or auditing) get attention. The higher security level the more sophisticated protection is required. But the higher levels also have more stringent assurance requirements. There must be more reason to believe that the system functions as designed.

The ITSEC also classifies the security of systems, so does the Common Criteria. A product or a system can be certified for a particular security class. The vendor asks an independent organisation to evaluate properties of a particular product/system and if this *Target of Evaluation* complies with the criteria, the label is granted. Although an obtained security label does not automatically imply that the product is secure, it helps in product categorisation and comparison.

In this chapter we categorise biometric systems according to the level of protection they offer. Our classification proposal divides systems into four levels. We first introduce the model of a biometric system. Then adjustable and/or optional parameters of biometric systems are discussed and at the end four security levels are described.

4.1 Modules of a biometric system

Any biometric system is basically made of the following components:

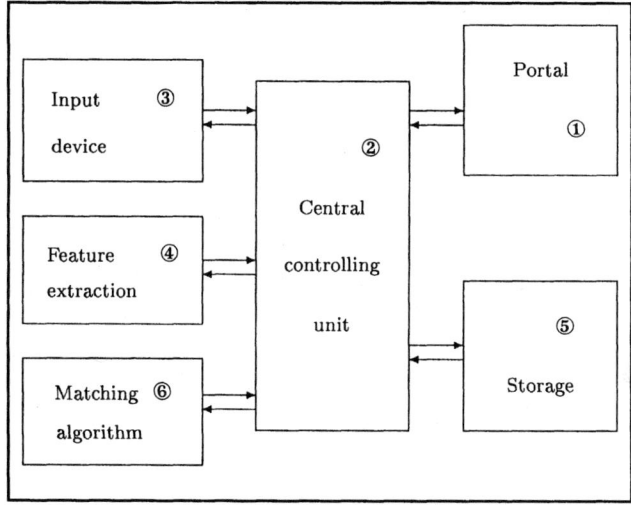

Figure 1. The model of a biometric system.

1 *Portal.* Its purpose is to protect some assets. An example of a portal is the gate at an entrance of a building. If the user has been successfully authenticated and is authorised to access an object then access is granted.

2 *Central controlling unit* receives the authentication request, controls the biometric authentication process and returns the result of user authentication.

3 *Input device.* The aim of the input device is biometric data acquisition. During the acquisition process user's liveness and quality of the sample may be verified.

4 *Feature extraction* module processes the biometric data. The output of the module is a set of extracted features suitable for the matching algorithm. During the feature extraction process the module may also evaluate quality of the input biometric data.

5 *Storage* of biometric templates. This will typically be some kind of a database. Biometric templates can also be stored on a user-held medium (e.g., smartcard). In that case a link between the user and her biometric template must exist (e.g., in the form of an attribute certificate).

6 The biometric *matching* algorithm compares the current biometric features with the stored template. The desired security threshold level may be a parameter of the matching process. In this case the result of the matching will be a yes/no answer. Otherwise a score representing the similarity between the template and the current biometric sample is returned. The central unit then makes the yes/no decision.

4.2   Parameters of biometric systems

What does it take for one biometric system to be more secure than another one? What are the differences among various systems?

Liveness testing: Incorporation of a liveness test makes an attack against the biometric system more difficult. There are various liveness tests offering various levels of protection. Most of the tests, however, can be easily cheated. A combination of multiple liveness tests can make the system more secure.

Tamper resistance: If the biometric system is not under constant human supervision it has to rely on tamper resistance. Without tamper resistance or supervision the system can be tampered with and forged/replied biometric data can be injected into the system.

Secure communication: Biometric system components can be either standalone and communicate with each other over an external insecure medium or can be coupled in a tamper-resistant box. The communication among modules within a tamper-resistant cover need not be secured, but the communication over an insecure line should be authenticated and encrypted.

Security threshold level: Lower false acceptance rate means higher level of security (and unfortunately, in most cases, also higher false rejection rate causing user frustration). A proper value must be set in accordance with goals of the biometric system.

Fall-back mode: In some systems the biometric authentication may be sufficient for the user authentication. In some systems an additional authentication method must be used and the biometric authentication is only a necessary part of user authentication. Successful authentication using this additional method may but need not be sufficient for user authentication.

4.3 Proposal of classification

Our proposal of classification divides biometric systems into four categories according to the level of security they offer. The higher security category the higher level of protection the system offers. Which level to choose depends heavily on the purpose of the biometric system, its threats and on available funds.

Level 1 – Very simple systems: Systems falling into this category are more or less very simple. They offer only restricted level of protection and can be easily cheated. Such systems have no liveness test incorporated and no part of the system has to be tamper-resistant. The communication among particular components need not be authenticated nor encrypted. Successful biometric authentication is sufficient means of authentication and after an unsuccessful biometric authentication some traditional authentication method is offered.

Such biometric systems are subject to easy attacks such as unplugging the biometric input device and injecting previously eavesdropped biometric data (because of no encryption or authentication), misuse of high false acceptance rate or faked trivial copies of biometric characteristics.

Level 2 – Simple systems: Biometric systems at level two require mutual authentication of particular components and encrypted communication. Still no liveness testing or tamper resistance is required. The biometric authentication is sufficient authentication. A traditional authentication method as a sufficient authentication method is offered only in the case of biometric system malfunction.

Systems on level two offer a certain level of security and still remain relatively cheap. Some of the easiest attacks are eliminated, but the systems still can be tampered with or cheated with faked biometric characteristics.

Level 3 – Intermediate systems: Level three systems already do have some kind of liveness test. Exposed components of the system (typically the biometric input device) must be guarded or tamper-resistant against moderate attacks. The communication must be authenticated and encrypted. The biometric authentication is sufficient, and the system never offers traditional authentication as a sufficient authentication method.

Such biometric systems will be able to resist moderate attacks. Advanced tampering methods or advanced faked biometric characteristics, however, will still be able to cheat the biometric systems.

Level 4 – Advanced systems: For systems of level four more than one advanced liveness test method are required. Exposed and unguarded components must be tamper-resistant. Such tamper resistance must be able to resist advanced tampering attacks. Communication among particular components (except within a tamper-resistant box) must be mutually authenticated and encrypted. Successful biometric authentication is necessary but not sufficient part of the user authentication. A supplemental traditional authentication method must be a necessary part of the authentication, too. Preferably multiple biometric techniques should be involved in the biometric authentication.

Biometric systems falling into the level four should be able to resist even professional and well-funded attacks. But nothing is bulletproof and designing a system resistant to (for example) very well funded attacks of intelligence services is rather difficult.

Table 1. Brief overview of classification proposal.

Level	Liveness	Tamper res.	Secure Comm.	Traditional auth method
1	no	no	no	sufficient/any time
2	no	no	yes	sufficient/malfunction
3	yes	moderate	yes	not sufficient
4	multiple	advanced	yes	not sufficient/required

5. CONCLUSIONS

Let us discuss where the use of biometric systems may be an advantage and where not. Biometrics are a great way of authenticating users. The user may be authenticated by a workstation during the logon, by a smart card to unlock the private key, by a voice verification system to confirm a bank transaction or by a physical access control system to open a door. All of these cases are typical and correct places where to deploy a biometric system.

Very promising are solutions where the cryptographic functions as well as the biometric matching, the feature extraction and the biometric sensor are all integrated in one (ideally also tamper-resistant) device. Such devices provide a very high protection of the secret/private key as the biometric data as well as the secret/private key will never have to leave the secure device.

We believe that biometric authentication is a good *additional* authentication method. Even cheap and simple biometric solutions can increase the overall system security if used *on top* of existing traditional authentication methods.

Biometrics can be used for dozens of applications outside the scope of computer security. Facial recognition systems are often deployed at frequently visited places to search for criminals. Fingerprint systems (AFIS) are used to find an offender according to trails left on the crime spot. Infrared thermographs can point out people under influence of various drugs (different drugs react in different ways). Biometric systems successfully used in non-authenticating applications may but also need not be successfully used in authenticating applications.

5.1   Where not to use biometrics?

Although good for user authentication, biometrics cannot be used to authenticate computers or messages. Biometric characteristics are not secret and therefore they cannot be used to sign messages or encrypt documents. If my fingerprint is not secret there is no sense in adding it to documents we have written. Anyone else could do the same. Cryptographic keys derived from biometric data are nonsense, too.

Remote biometric authentication is not trivial at all. The assumption that anyone who can provide my fingerprint can also use my bank account in the homebanking application is not a good idea. Remote biometric authentication requires a trusted biometric sensor. Will a bank trust your home biometric sensor to be sufficiently tamper resistant and provide trustworthy liveness test? Although remote biometric authentication may work in the theory, few (if any) current devices are trustworthy enough to be used for remote biometric authentication.

While using biometrics as an additional authentication method does not weaken the security of the whole system (if users do not rely on the biometric component so much to ignore the traditional authentication method, e.g., by using simple passwords), replacing an existing system with a biometric one may be more risky. Users as well as administrators and system engineers tend to overestimate security properties of biometric systems; such a decision must be based on and confirmed by a risk analysis. Particularly, reviewing the process of the biometric data capture and transfer is very important. Sometimes biometric authentication systems replace traditional authentication systems not because of higher security but because of higher comfort and ease of use.

False rejects – the unpleasant property of biometric systems causing authorised users to be rejected – may prevent biometric systems to spread into some specific applications, where inability of a user to authenticate herself (and run an action) may imply serious problems.

Few basic conclusions at the very end:

* *Different biometric samples of the same person will never be same.*

* *Biometric systems make errors.*

* *Biometric data are not secret.*

* *The role of the input device is crucial, and this device must be trusted or well secured.*

* *The biometric system should check user's liveness.*

* *Biometrics are good for user authentication. They cannot be used to authenticate data or computers.*

Notes

1. There are two kinds of errors that biometric systems do: *false rejection* occurs when a legitimate user is rejected and *false acceptance* occurs when an impostor is accepted as a legitimate user. The number of false rejections/false acceptances is usually expressed as a percentage from the total number of authorised/unauthorised access attempts. *The equal error rate (ERR)* is the point where FAR and FRR are equal. The ERR value as such does not have any practical use, but it can be used as indicator of the biometric system accuracy.

2. Both the FAR and FRR are functions of the threshold value and can be traded off, but the set of usable threshold values is limited. For example a system with the ERR of 1% may be set to operate at the FAR of 0.01%, but this would imply the FRR to jump over 90 or 95%, which would make system unusable.

3. The FTE rate is estimated as 2% for fingerprint based systems and 1% for iris based systems. Real values of the FTE rate are dependent on the input device model, the enrolment policy and the user population.

References

[1] Common Criteria for Information Technology Security Evaluation, v 2.1, 1999.

[2] Department of Defense (1985). Trusted Computer System Evaluation Criteria.

[3] Jain, A., Bolle, R. and Pankanti S. (1999). *BIOMETRICS: Personal Identification in Networked Society*. Kluwer Academic Publishers.

[4] National Institute of Standards and Technology (1994 and 2001). *Security Requirements for Cryptographic Modules, FIPS PUB 140-1/2.*

[5] Newham, E. (1995). *The biometric report.* SBJ Services.

[6] Matyáš, V., Říha, Z. (2000). *Biometric Authentication Systems.* Technical report. http://www.ecom-monitor.com/papers/biometricsTR2000.pdf.

[7] Mansfield, T. (2001) *Biometric Product Testing – Final Report,* National Physical Laboratory, 2001, http://www.npl.co.uk/.

AUTOMATIC AUTHENTICATION BASED ON THE AUSTRIAN CITIZEN CARD
A Reference Implementation

Arno Hollosi, Udo Payer, Reinhard Posch
Institute for Applied Information Processing and Communications

Abstract: The concept of the Austrian citizen card enables the generation of electronic signatures, provides mechanisms to establish confidential communication channels, and supports features for user authentication in public services. This document specifies a mechanism —*based on the Austrian citizen card*— and a trustworthy component —*called security layer*— to fulfil all requirements for authentication processes, suitable for electronic administrations. The additional trustworthy component (security-layer) forms the interface between diverse applications and the smart card (citizen card). But this layer also offers features which can be used very efficiently in conjunction with certificate-based user authentication. Depending on the used technology, three different levels (qualities) of user authentication can be realized. In the following, a short introduction is given to the concept of the Austrian citizen card followed by common descriptions of three mechanisms suitable for usage in the environment of public services.

Key words: citizen card, smart card, identity token, peer-entity authentication, authorization

1. INTRODUCTION

The concept Austrian citizen card is based on smart cards, which are able to generate secure electronic signatures. Today, traditional administrative requests are tightly coupled to the conventional signature of the concerned citizen. Therefore, electronic administration procedures have to offer equivalent possibilities - even in the case of electronic attachments.

The Austrian signature law and signature order form the legal basis of generating so called secure electronic signatures[2]. This enables citizens to enter the electronic administration without prior personal registration. The citizen card concept intends to offer public administration procedures which can be modelled efficiently and economically. This presupposes that electronic administration applications can be automated to a large extent. It also implies that the underlying infrastructure supports such mechanisms.

From all of its diverse forms of appearances, the citizen card concept will be based primarily on the social insurance card, since eight million citizens are under social security and will receive their own social insurance card. Mixing the citizen card concept with mechanisms for social security services is unobjectionable, since different cryptographic mechanisms are in use. An endangerment by cross-references of different ranges is impossible in each case. But other cards (identity card, bank cards, etc.) will be applicable in the citizen card concept as well.

Apart from generation of electronic signatures, the citizen card concept provides the possibility to store additional data elements in so called "information boxes". For example, it can be very useful to store certificates or other information on the card which makes online access to these resources unnecessary. Information boxes can also be used to store electronic documents. As the memory on smart cards is quite limited, it is also thinkable to store references to a repository instead of the documents themselves. Therefore, it would become possible to store documents in arbitrary places – protected by authentication based on the card. Thus, the card owner has full control over content and volume of these information boxes.

Due to the multitude of people involved and their different requirements, the citizen card concept intends to provide roles and mandates. Attribute certificates and other methods of IT-security will be used in order to be able to realise roles and mandates technically.

Public administration entrances have a special need of privacy. This document describes solutions of integrating cryptographic mechanisms exemplarily which are suitable for user authentication in electronic administration. According to different approaches, three stages (qualities) of user authentication can be defined. A description of these solutions is given in section 4.

But before turning to technical details of the implementation, some fundamentals and requirements should be discussed first:

---

[2] Note that "secure electronic signature" is the term used by Austrian law. In the European directive these signatures are called "qualified electronic signatures".

2.   FUNDAMENTAL PRINCIPLES

This part discusses some common authentication mechanisms and points out advantages and disadvantages of current solutions. Moreover, general requirements of e-Government solutions are specified.

Fortunately, the Austrian citizen card will be able to support ECC mechanisms, which is undisputed in connection with smart cards. Therefore, a short introduction into digital signatures – based on ECC mechanisms – is given first.

2.1   ECDSA and the Austrian Citizen Card

The Elliptic Curve Digital Signature Algorithm (ECDSA) [3] is the elliptic curve analogue of the well-known Digital Signature Algorithm (DSA). It was approved in 1999 as an ANSI standard [4], and was accepted in 2000 as IEEE and NIST standards. Moreover, ECDSA was also accepted in 1998 as an ISO standard.

Unlike the ordinary discrete logarithm problem and the integer factorisation problem, no sub exponential-time algorithm is known for the elliptic curve discrete logarithm problem. For this reason, the strength-per-key-bit is substantially greater in an algorithm that uses elliptic curves.

ECC systems over prime fields of at least 161 bits (or 188 bits if one prefers to be more careful) are sufficient until year 2020 [6]. As the social insurance card supports ECDSA, it is clear that these mechanisms will play a major role in the concept of the Austrian Citizen Card.

2.2   Common Solutions

User authentication is a term which is used in a very broad sense. By itself it has no other meaning than the guarantee that users are who they claim to be. User authentication in web-based applications is therefore a much-discussed topic followed by a great number of different solutions:

Almost all browsers support **basic authentication**. When entering a realm, a standard popup window appears on the screen, asking for username and password. The realm value should be considered an opaque string, which can only be compared for equality with other realms on that server. The server will service the request only if it can validate the user-ID and password for the protection space of the Request-URI. The major drawback of the basic authentication scheme is that it is relatively simple for eavesdroppers to spy out the password since it is transmitted as plain text. [10]

An alternative authentication scheme known as **digest authentication** remedies this weakness through the use of crytographic hashes, usually the MD5 message digest algorithm defined in [7].

Now, while taking username and password, running them through MD5 (as you do with base64 for basic authentication), and sending the result to the server, a potential eavesdropper could record the hashed username and password to initiate a replay attack.

To securely prevent replay attacks, a more sophisticated procedure is obviously necessary - the **digest access authentication** scheme:

The main difference between digest- and digest access authentication is the use of a nonce to prevent replay attacks. But this scheme provides no encryption of message content. The intent is simply to create access authentication methods to avoid the most serious flaws of basic authentication without using SSL [10]

Cookie authentication makes use of functionality at the scripting level to provide user authentication. Cookies can store usernames and passwords independently of locally stored users - this requires that browsers support cookies and that cookies are enabled. Thus, it is not practicable to use this mechanism on publicly accessible computers, such as public portals or university environments.

Beyond standard HTTP authentication schemes, there are a couple of authentication mechanisms, either based on **sending encrypted user information**, or **using digital certificates**.

The **ISAPI** (Internet Server Application Programmer Interface) provides low-level access to the entire web server request and event chain. Because of this it can intercept requests before the web server handles them and provides the greatest authentication flexibility. ISAPI interception plays a major role in some of the proposed solutions.

2.3   General Conditions for e-Government Solutions

This section discusses additional demands caused by heterogeneous environments (web browsers and clients) that are expected in standard e-Government solutions.

Simple processes: The number of used technologies should be limited. Thus, a rapid conversion and adjustment of new applications can be accomplished easily. This also implies openness for all given systems. The people involved (developers of applications for electronic administration) should have to deal with a minimum number of simple structures. These simple structures and interfaces should be simple to learn and deployable without large efforts.

Open Interfaces: Core interfaces should be in the public domain and be built upon international, non-proprietary standards and technologies. This avoids vendor lock-in. Also, these interface should cleanly separate key players and their legal liabilities. The interface to the trustworthy component (security-layer) is implemented as an open interface based on TCP/IP, HTTP, and XML.

Authorization and certificates: Authorization should be based on end-to-end mechanisms. Public access or privately held facilities are not allowed to play a role in the course of user authorization. Therefore, certificates and electronic signatures are well suited for dynamic user authorization. Directory services can be used to share certificates and attribute certificates. These components are forming the essential mechanisms to integrate roles and positions into user authorization. Since usage of certificates in authentication processes is anything but simple, the security-layer is used to simplify all processes by offering an elementary interface.

Authentication: User authentication should be realized by using electronic signatures. Declaration of intention to sign an authentication request should always require the entering of a PIN. For less important authentication scenarios it is also conceivable to use other mechanisms.

Authentication of users is usually limited to the authentication of physical persons. If additional characteristics have to be proven, the best solution would be to fall back on conventional, paper-based mechanisms suitable for conventional administrative authorities. These mechanisms can be based on resident certificates, birth certificates etc.

In a similar way, it can be necessary to enclose these additional certificates in an electronic process. A suitable method to deal with these documents is to store their content in XML structures.

Administration officers – working with electronic documents – have to be authenticated as well. In principle, the same mechanisms as used for authentication of citizens can be applied. Characteristics and roles of administration officers can be managed by attribute certificates. In the case of accepting a role, the officer has to be identified in the context of the application and not in the context of access services.

Single Sign On: Single Sign On is suitable and required for processes in a distributed application framework. Single Sign On can also be realized by using electronic signatures. The main intention is to avoid multiple identifications to act in different roles or deal with different applications following a single authentication. Multiple authentications in the course of a single session are not necessary and desirable. Once authenticated, the user remains authenticated as long as the user does not leave the same security realm.

Confidentiality: A certain degree of confidentiality is also required by some processes. This can be realised by using encryption mechanisms. Methods and key lengths have to be chosen in such a way that claimed levels of confidentiality can be guaranteed.

Beyond this, we have to guarantee secure end-to-end connections. Smart cards are well suited for this purpose, as their private keys are protected from disclosure and they can be used for electronic signatures to prevent unauthorized access to confidential information.

Identification: Austria has strict privacy protection laws. In context of the citizen card concept every citizen is assigned a personal ID number which is essential in order to easily and accurately identify a person. However, law forbids that this number is stored in databases. This gordic knot is solved by using one-way functions to transform the PID into a context dependant process identifier which can be stored in databases. Basically, the transformation takes the PID and the application name as input and produces the process identifier (e.g. by using a hash function). Thus a person has different identifiers in different applications. Note that during the authentication process the PID is transmitted, so that the server can derive the context dependant ID.

3.   ACCESSING THE CITIZEN CARD TRHOUGH THE SECURITY-LAYER INTERFACE

To access Austrian citizen cards a trustworthy interface is required. The security-layer represents this interface, which is one of the central components in the proposed authentication process.

The security-layer acts as an interface between the browser and the citizen card and can be used to send signature- and verification requests to the citizen card. Moreover, this layer also provides commands for reading and writing data to information boxes (short: info boxes).

Connecting browser applications to the security-layer is quite simple since two different interfaces are supported. Beside a standard TCP/IP and socket-based interface, it is also possible to pass over XML requests [13] in the form of HTTP POSTs [8]. Requests like signing or verifying XML structures can be sent to the security-layer – which is processing this request – and retrieved by the requestor or can be forwarded to a given URL.

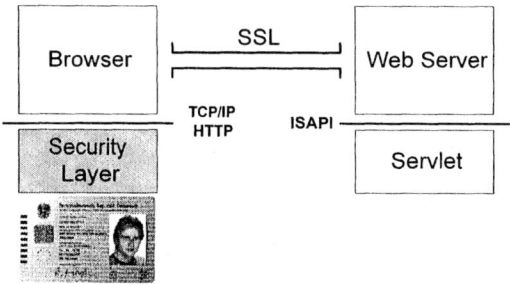

Figure 1: Security-layer offers two interfaces: TCP/IP and HTTP

Another key function is the ability to transform the XML response into HTML using XSLT stylesheets loaded from the server. This HTML, which may contain scripts as well, is then sent to the browser. This mechanism is playing a major role in one of the proposed authentication mechanism, unless further client components are installed.

Just for completeness we have to note, that neither socked based nor HTTP based bindings are inviolable against authentication and confidentiality on the interface to the browser. Thus, a limitation to local host access is mandatory and self-evident. Figure 1 depicts that just local browsers are authorized to bind the security-layer.

4. AUTOMATIC USER AUTHENTICATION

Online applications within the range of the public administration require different stages of security. This also applies to the level of authentication, since applications may exist which do not need qualified certificates[3] [14]. As a function of the authentication level, three stages can be defined:

1. Safe for normal operations
2. Safe within a trustworthy infrastructure
3. Technical end-to-end safety

A common public application requires just a minimum of confidentiality. These services can be realized by a simple server-authenticated SSL connection.

If a trustworthy infrastructure is to be achieved, the use of active components is imperative. These active components can either be loaded

---

[3] Qualified certificate and advanced electronic signatures are common terms in the European Community framework for electronic signatures.

from a trustworthy side (whenever they are needed) or can be installed on the client's host.

The technical realization of trustworthy end-to-end channels can either be based on certificates, or mechanisms based on advanced electronic signatures, created by using qualified certificates.

These security requirements have to be granted in ordinary technical environments. It can be the fact that some of these requirements have to deal with (1) different client certificates, since no common certificate structures exists. Secondly, the introduction of trustworthy active components can be realized by using the (2) concept of the citizen card in combination with the security-layer. And finally, some of the Austrian citizen cards are based on (3) elliptic curve mechanisms. Thus, ECC mechanisms have to be supported.

4.1   Level I

Level I fulfils only rudimentary demands on the quality of the communication channel. Basically level I guarantees confidential, one-side authenticated communication. Respective requirements can be achieved by simple server-authenticated SSL or TLS connections. In the event of accessing a sensitive realm, secure communication has to be requested by using SSL or TLS. Beyond that, guidelines have to exist, enforcing a minimum key length of at least 100 bits. Apart from well-known problems with server-authenticated SSL connections, there is a theoretical possibility of Man-in-the-Middle attacks.

4.2   Level II

Level II makes higher demands on the trustability of the communication channel. It also requires user authentication. This level is based on signing and verifying authentication information. In doing so, authentication is based on a mutual agreement, very similar to the X509 strong 2-way authentication protocol.

Right after an authentication request (0), the server has to create and sign a security token (1) – consisting of a timestamp (or nonce), the session ID, and the servers IP address or URL – and has to send this signed security token to the requesting client. The client has to verify the server's signature – has to extract timestamp and URL and has to compose a unique authentication block (2). Uniqueness of the authentication block is based on time stamps and the signature, which is created by using a qualified certificate. This security token has to be sent to the URL which was included in the server's security token. Finally the server has to check the client signature, derives the client's identity information, and registers the user to the selected application.

```
0. C→S:  Auth.Req.
1. C←S:  certS, SS(tS, URLS, SID)
2. C→S:  certC, SC(tS, URLS, SID, IDL)
```

IDL = SBH(KPC1, KPC2, C) ... Identity Link
KPC1 ... public key
KPC2 ... public key of a qualified certificate
C   ... personal information (name, date of birth, personal ID)
SBH() ... signed by the public authority
SID   ... Session ID
tS   ... time stamp or nonce

Figure 2 describes the process of building and signing the security token (1) with the subsequent generation of an authentication block (2), figurative:

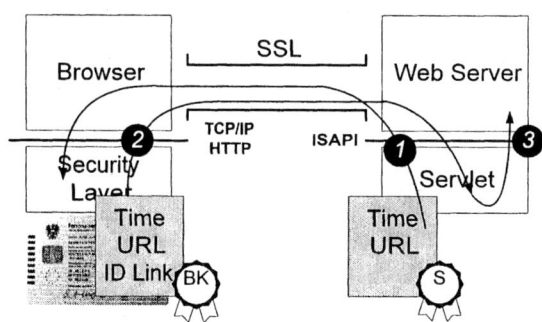

Figure 2: Signing and verifying SCT and authentication block

Level II can be overwhelmed if attackers are in the possession of the server certificate (S) issued to administration servers. However, this is unlikely to happen. A more serious issue arises, if users do not verify the server certificate, which unfortunately is common practice. Assume that a user can be tricked into accessing an arbitrary server (e.g. having a similar URL) that holds a server certificate S'. S' might be issued from a certification authority trusted by the user's browser and thus no security warning appears on the screen. This arbitrary server can then act as man-in-the-middle (until the user verifies the certificate and detects the fraud).

To clarify components and services, section 4.2.1 to 4.2.4 will specify involved data structures and mechanisms in more details.

4.2.1   Security Token

Right after accessing restricted pages, the server has to generate a time stamp. A time frame is started, within the client has to respond by sending the authentication block. The timestamp together with the session ID, IP Address or server's URL forms the security token, which has to be signed by the server and has to be sent to the client. All this has to occur in already established server authenticated SSL connections.

4.2.2   Authentication Block

Initiated by the identification request, the client has to verify the server signature by using the server certificate. If this signature is valid, the client has to check the session ID, and has to memorize the server URL. The next step is to fetch the identity link from the citizen card, which is permanently stored and can be secured by cryptographic mechanisms in one of the citizen cards data bags (info box). These data bags can also be used to store electronic documents in the form of XML structures, certificates, or attribute certificates.

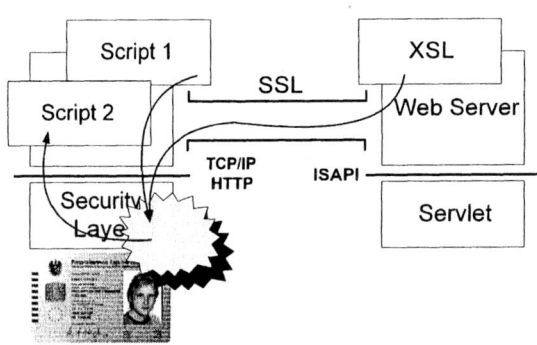

Figure 3. Transforming style sheets (XSL)

Reading the info box is done by a script, which is followed by submitting a form, containing a stylesheet URL. The security-layer is loading the specified stylesheet (XSLT) from the server and transforms it into an HTML page containing a script, which is executed – next to the previous one – by the browser.

This subsequent script is used to compile the authentication block. To be able to generate variable scripts – since scripts have to be generated on a per connection basis – another active component is used at the server side. This component supplies the client with downloadable and modified scripts and pages.

The completed and signed authentication block is sent back to the active server component, where its signature is verified.

4.2.3   User Registration

Right after verification of the authentication block, the active component is extracting the personal ID number (PID) from the identity link to calculate the process identifier. As discussed in section 2.3, one-way functions have to be used to derive a unique process identifier from the unique PID. The derived process identifier can be used as username (and password,) which are used to register users to e-Government applications.

As long as a user remains within the same realm, the user remains registered.

4.2.4   Proxy Services

If the registration of users to applications is based on simple authentication processes, access to protected pages has to be limited to the server. Once registered to the server application, the servlet has to mark this connection as an authenticated connection and has to act as a proxy for the authenticated client.

4.3   Level III

The policy of level III is in principle based on the same mechanisms as describe in section 4.2. The only difference is the use of the SSL server certificate, which is tightly bound to the used SSL connection, to form the security token (SCT).

After generating and sending the SCT to the client, the client can extract the server certificate from the corresponding SSL connection and can use this certificate to verify the SCT signature. After successfully verifying the SCT, a trustworthy server authentication on the client side can be assumed.

This mechanism can successfully prevent man-in-the-middle attacks and spoofs relying on "lazy" users not verifying certificates. But realization of level III authentication requires the integration of a further trustworthy active component at the client, to obtain the required server certificate.

Browser Helper Objects [1] can be used to get access to any running instance of an Internet Explorer by attaching itself to every new instance. By using this feature, it is possible to gain easy access to the object model as well as to receive all events coming from the browser. A BHO is therefore an excellent device to intercept and modify HTTP data streams, and well suited for level III implementations.

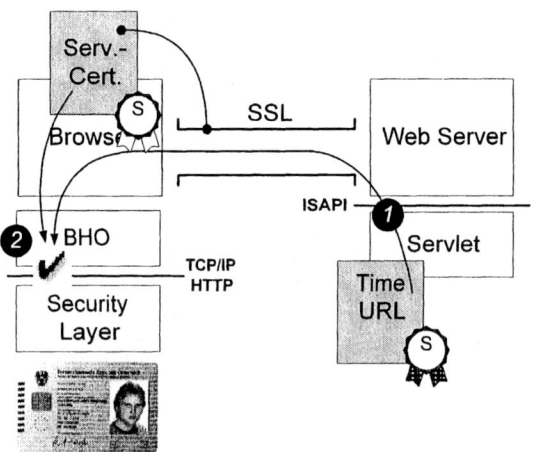

Figure 4: Requirement of a trustworthy instance (BHO) to obtain the SSL certificate

Once the signature of SCT was verified, an authentication block is built on the same way as described in layer II.

5.   CONCLUSION

The concept "Austrian citizen card" takes into account mechanisms for electronic signatures, describes interactions between the citizen card and e-Government applications, which may make use of preliminary portals or market places. Substantial characteristics of these mechanisms are directory services, certificates and their attributes. This paper describes some features of the concept of the Austrian citizen card (which are of interest in context of authentication) and how this concept can be used for user authentication processes using standard web technologies. It should be noted that the discussed procedures, applications, and application structures are suitable not only for e-Government but for generic web applications as well. Furthermore, we discussed how the available underlying technologies allow different qualities (levels) of user authentication. From the offered procedures, application developers are free to choose a suitable one, which either is well suited for a certain application or offers a maximum of security.

REFERENCES

[1] Esposito: Browser Helper Object: "The Browser the Way You Want IT", Microsoft Corporation, January 1999.

[2] Certicom, "Elliptic Curve Cryptosystem for Smart cards", Certicom White Paper, 05/1998.

[3] ANSI X9.62, "Public Key Cryptography for the Financial Services Industry": The Elliptic Curve Digital Signature Algorithm (ECDSA), 1999.

[4] American National Standard Institute, "Public Key Cryptography for the Financial Services Industry: The Elliptic Curve Digital Signature Algorithm (ECDSA)", ANSI X9.62-1998, 1998.

[5] Karlinger: "XML Electronic Signatures Application according to the international standard XML Signature Syntax and Processing", CMS 2001 Darmstadt, Germany, 2001.

[6] Lenstra: "Selecting Cryptographic Key Size", The Journal of the International Association for Cryptology Research, Vol.14 Number 4, 2001.

[7] R. Rivest: "The MD5 Message-Digest Algorithm", RFC1321, April 1992.

[8] L. Daigle, D. van Gulik, R. Iannella, P. Faltstrom: "URN Namespace Definition Mechanisms", RFC2611, June 1999.

[9] Gettys, Mogul, Frystyk, Masinter, Leach, and Berners-Lee: "Hypertext Transfer Protocol HTTP/1.1",RFC2616, June 1999.

[10]Franks, Hallam-Baker, Hostetler, Lawrence, Leach, Luotonen, and Stewart: "Basic and Digest Access Authentication", RFC2617, June 1999.

[11]Reagle: "XML Signature Requirements", RFC2807, July 2000.

[12]Reagle, Eastlake, Solo: "XML Signature Syntax and processing", RFC3075, March 2001.

[13]IETF W3C: "XML-Signature Syntax and Processing".

[14]The European Parliament and the Council of the European Union: "DIRECTIVE 1999/93/EC OF THE EUROPEAN PARLIAMENT AND OF THE COUNCIL", Official Journal of the European Communities, Article5, December 1999

AN OPEN INTERFACE ENABLING SECURE E-GOVERNMENT
The Approach Followed with the Austrian Citizen Card

Arno Hollosi[1], Herbert Leitold[2], Reinhard Posch[1]
[1] *Chief Information Office (CIO) Austria,*
    {Arno.Hollosi, Reinhard.Posch}@cio.gv.at
[2] *Institute for Applied Information Processing and Communications (IAIK),*
    Graz University of Technology, Herbert.Leitold@iaik.at

Abstract: When encouraging citizens to approach public administrations by electronic means in order to improve the public services and to avoid costly media transitions from paper-based applies to IT-supported back-office applications, authorities and implementers need to be in particular cautious in two aspects: On the one hand, security is an indispensable guiding principle for concerns of legal certainty, identification and authentication requirements, confidentiality and data protection aspects, and certainly security is needed to achieve broad user acceptance. Electronic signatures based on smartcards represent a state-of-the-art in supporting several of these security requirements. On the other hand, the concepts followed need to be technology-neutral to a large extent to both remain open for future or emerging technologies that may mature to meet these security requirements as well and to avoid discrimination against particular solutions. Otherwise inclusion of upcoming solutions may well turn out a costly experience. In this paper the approaches followed with the Austrian citizen card are discussed – an ambitious project that aims at deploying e-Government on the large scale. By means of an open interface the authorities specify the requirements arising out of the applications in the administrative bodies. This allows the authorities to launch the development of applications based on well-defined interfaces, but not mandating a certain technological instantiation such as a social security card, public identity cards, or private-sector-borne signature cards such as banking cards. By taking up and implementing the interface specification an open market is stimulated that paves the way to a public-private partnerships. The paper gives the rationale of choosing the open interface approach and discusses its actual implementation – the so-called security layer – in detail.

Key words: electronic signatures, open interfaces, citizen card, identity card

1. INTRODUCTION

In a summit on 20th November 2000 the Austrian federal government unanimously decided to employ chip-card technology to simplify approaching public authorities for citizens. This decision resulted in a momentum towards providing the required infrastructures and applications. For instance, the Austrian social security card that will be rolled out to each citizen end of 2002 has been enhanced by electronic signature functionality fulfilling the Austrian signature law [1] and the Austrian signature order [2] which is required for identification and data origin authentication purposes. Similarly, a public identity card has been defined based on smart-card technology. The legal framework has been prepared for e-government by permitting derived and encrypted (one-way hash functions of) public registration numbers as a means of identification [3] or enabling electronic delivery of official notifications [4]. Moreover, numerous pilots and applications have been launched or accommodated for electronic signatures by the various federal ministries such as "finance office online" enabling online tax declarations, "help@gv" an online information platform, or electronic confirmation of payment as a means of prove that administrative fees have been remitted.

However, from an holistic perspective it seems shortsighted to confide the success of e-Government solely on a multitude of technologies or applications, it is apparent that proceedings of authority's processes are interweaved – involving different administrative units and applications – and thus generalized approaches appear as the road to follow. Moreover, two aspects are imperative when considering deploying information and communications technologies (ICT) end-to-end between the citizen and the administration – **security** and **openness:**

– Security measures are required in numerous aspects: Public authority's proceedings require identification of the parties involved in many cases, data origin authentication and requirements of writing are given, or data protection laws are to be followed. In particular, public administrations can not take residual risks to an extent as they are taken in e-commerce scenarios in many cases.

– Regarding openness, applications and infrastructures for e-Government are considered a long-term investment. While electronic signatures, public key infrastructures (PKI), and smartcards are considered appropriate in fulfilling the security requirements listed above, it is well conceivable that off-the-shelf technologies such as cell phones or personal digital assistants (PDAs) will mature in terms of security and will compete with the conventional '*smartcard and PC*' combination. Adapting each application to new technologies as they show up is costly.

However, not including appropriate technologies results in the dilemma of being discriminatory with respect to future technologies. Thus, a technology-neutral road needs to be followed that allows the citizen to employ the technology of choice and still keeps the resulting requirements on the administration's application in a reasonable and manageable size.

The approach followed with the Austrian e-Government initiatives was to base the proceeding on a coordinating initiative that resulted in a consolidated view on the requirements from the administration's perspective [5]. This resulted in a list of general demands on the citizen's security token which may show up in a variety of appearances. We refer to the totality of appearances as the "Austrian citizen card" – one might suspect that this mandates smartcards, but in fact a number of alternative technologies are conceivable. However, in its initial phase smartcard systems such as the social security card will represent the vast majority.

The general demands on the functionality of the citizen card are basically (1) creation of so-called secure electronic signatures[4] that are equivalent to handwritten signatures and fulfill the requirement of writing according to the Austrian Signature law [1], (2) a second key pair which can be used for peer entity authentication or for establishing session certificates and session keys for securing the communication, and (3) so-called info-boxes that serve as containers for storing data such as certificates, identifiers with or without access control, or other data. Based on these requirements an open interface has been defined – the so-called security-layer. The security-layer offers a transmission control protocol, internet protocol (TCP/IP) communication interface and extensible markup language (XML) as the basic format of its protocol data units. This interface fulfils the requirement of the e-Government application in a generalized fashion, as well as gives the market a maximum flexibility for joining in the Austrian citizen card project. It therefore is discussed in detail in this paper.

The remainder of the paper is structured, as follows: In section 2 the legal and normative framework in which e-Government operates is discussed. This is mainly the signature law and the signature order. As the European common market asks for cross-border interoperability of national solutions, the European dimension is addressed in terms of the EU Signature Directive [6] and the European standardization efforts carried out by the European Telecommunications Standards Institute (ETSI) and the European

---

[4] Note, that the term 'qualified electronic signature' is also common for an electronic signature that is equivalent to a handwritten signature in legal terms, such as in the German signature law or in the standards developed in the European Electronic Signature Standardization Initiative (EESSI). However, the notion 'secure electronic signature' is used here, as this is the terminology used in the Austrian Signature law [1].

Committee for Standardization (CEN). Section 3 continues by giving details on general requirements when deploying e-Government. This gives an extended view to the basic requirements **security** and **openness** that have been sketched above. The security-layer as an open interface for e-Government applications is discussed in detail in section 4. Finally, conclusions are drawn.

2.   LEGAL AND NORMATIVE FRAMEWORK

In order to contribute to the legal recognition of electronic signatures, the EU Electronic Signature Directive established a framework that required the Member States to adopt its measures and bring the corresponding national laws into force by July 2001 [6]. In its article 5.1 the Directive defined that an electronic signature is admissible as evidence in legal proceedings and satisfies the legal requirements of a signature in relation to electronic data in the same manner as a handwritten signature in relation to paper-based data, if the electronic signature fulfils certain requirements – we refer to such a electronic signature as a 'secure electronic signature'. These requirements for a secure electronic signature are mainly that:
- it is created by a so-called secure signature-creation device (SSCD) which basically is the device getting in touch with the signature-creation data – the subscriber's private key spoken in common PKI terminology.
- it is based on a qualified certificate – the electronic attestation linking signature-verification data (SVD) – the subscriber's public key in PKI terms – to a person. In order to be qualified the certificate and the certification service provider (CSP) needs to fulfill certain requirements.

The requirements stated above are specified in the Annexes of the Directive, where Annex I defines the contents of a qualified certificate, Annex II defines the requirements for CSPs that issue qualified certificates, and Annex III addresses the SSCD. In addition Annex IV states recommendations on secure signature verification.

The Austrian signature law [1] has been put in force in January 2000 and the signature order [2] gives a greater detail in particular regarding technical aspects such as cryptographic key sizes. The signature law closely follows the European Directive. However, while the Directive limits mandatory certification of conformity with its provision by designated bodies to the SSCD, the Austrian signature law requires that the technical components and procedures for generating secure electronic signatures must be verified and that the conformance must be certified by a confirmation body. While one might assume this a minor difference to the provisions of the Directive, it yet has an important influence to the system design that is discussed in this paper: Note, that the notion of 'technical components for generating the secure electronic signature' as defined in the Austrian signature law is more general than limiting the scope to the SSCD, as the component for viewing the data to be signed (DTBS), or the component for authenticating the signatory such as a personal identification number (PIN) pad are included. This is further discussed in section 3.

In order to stimulate interoperable solutions, EESSI entrusted CEN and ETSI to develop standards that support the EU Signature Directive [7]. The main provisions that have been developed within EESSI[5] and which basically fulfill the requirements of the Austrian signature order, are:
- For SSCDs, Common Criteria (CC) [8] Protection Profiles – the SSCD-PPs [9] – have been defined.
- Based on cryptographic message syntax (CMS) [10] electronic signature formats have been developed [11], or XML signatures [12] have been specified.
- For CSPs issuing qualified certificates, policy requirements have been defined [13], as well as protection profiles for hardware security modules [14].
- The cryptographic algorithms and its parameters have been specified in a separate document [15]. Signature suites based on Rivest, Shamir Adleman (RSA) [16] and digital signature standards (DSS) [17] with a minimum key size of 1020 bit or signature suites based on elliptic curve cryptography (ECC) [18] [19] with a minimum of 160 bit keys have been specified.

Based on this legal and normative framework, we continue in the following section by discussing what basic requirements can be derived. In particular, organizational aspects and issues regarding forward compatibility are addressed.

3. BASIC REQUIREMENTS OF E-GOVERNMENT

When deploying e-Government on a large scale it is essential to focus on the impact on current processes and on the long-term effect the newly created infrastructure has on organizations. It seems vital that the following principles govern the design of an e-Government infrastructure:
- *Avoidance of vendor lock-in:* The transition of paper-based processes to electronic processes should not create new dependencies or monopolistic situations. Thus the interfaces of the core infrastructure have to be in the

---

[5] Note that these standards are not yet recognized European standards that in lieu is to be adopted by the member states.

public domain or need to be federal property. This ensures that the administration is never unconditionally bound to a single vendor.

- *Modular design:* The sophisticated electronic processes and transactions, the multitude of participating parties, and the legal responsibilities demand a clear division into atomic components with unambiguously defined interfaces, functions, and responsibilities.
- *Technology neutral design:* Advancements in technology are happening at an ever increasing rate. In order not to be caught in the vicious cycle of perpetual updates or shutting out emerging technologies, the design of the core infrastructure has to be ignorant of underlying technologies to the greatest extend possible.
- *Security and trust:* The basic element providing identification and authentication is the electronic signature. Thus for both the application filed by the citizen and the administration generating official notifications, secure electronic signatures provide the legal certainty. In addition, the PKI-based infrastructure builds a basis for additional certificates for confidentiality.

Taking these design principles into account it is evident that there is need for a high-level interface to the citizen's security token (or smart card). This interface should describe the security token not in terms of algorithms, but in terms of functionality. By encapsulating the functions and responsibilities in a clearly defined component which is accessed through a single interface to security tokens a flexibly integration into the e-Government structure is provided.

Austria's citizen card is therefore a 'requirements profile' for security tokens rather than a specification of smart card features. The requirements of the Austrian citizen card profile are based on the requirements created by e-Government processes. Firstly, there is need for creating secure electronic signatures to be able to submit applications in electronic form. As the signature law confines the way in which secure electronic signatures can be used, the need for a second certified key pair for entity authentication or similar areas arises.

A major advantage of the security-layer concept is that the security-relevant components are viewed as a single entity by the application. I.e., the SSCD, the document viewer component, and the PIN pad are under control of the developer. Replacing components, such as replacing a PIN pad by biometric sensors is transparent to the application. By following the security-layer concept as a requirement catalogue, the market thus gains maximum flexibility in developing solutions. The details of the security layer are given in the following section.

4. SECURITY LAYER – AN OPEN INTERFACE

In this section we are taking a closer look at the security-layer interface. The security-layer is the interface offered by the so-called security-capsule – the entity containing the citizen card and implementing its immediate environment which applications communicate with. The distinction between the interface and entities implementing its functionality is important, as an open interface can be defined technology-neutral to a large extent. Thus, in a public-private partnership a non-discriminatory approach can be followed where the public authorities define the interface – the security-layer – and the industry is free to implement the citizen card – the security-capsule.

As stated in the previous section, among the major goals of the security-layer concept are forward compatibility and independence from underlying technologies. Therefore, the interface uses TCP/IP connections as communication channel and XML encoding for the communicated protocol elements. This choice appears reasonable, as TCP/IP stacks are available even for the smallest devices and for fringe operating systems. The choice of XML is straight-forward, as it is aimed that documents to be signed are primarily XML documents, and that the applications utilizing the security-capsule will be XML-aware in most cases. Furthermore, XML is human-readable and can be easily parsed; the verbosity of XML is a negligible drawback.

The security-layer uses a straight-forward request/response protocol scheme. The application opens a connection to the capsule, sends its request, and waits for the answer. The capsule receives the request, processes it, sends a response, and closes the connection. The protocol and its XML encoded data can utilize different transport layers. Among the layers currently specified are plain TCP/IP, hypertext transfer protocol (HTTP), transport layer security (TLS), and HTTP over secure socket layer (HTTPS). Simple object access protocol (SOAP) and XML remote procedure calls (XML-RPC) are currently being investigated.

4.1 Protocol, Functions, and Commands

The security-layer provides the following high-level functions to applications:
- signing documents according to CMS [10] or XMLDSIG [20]
- verifying CMS or XMLDSIG signed documents
- storing and retrieving data
- utility functions such as creating a session certificate or creating a symmetric session key based on Diffie-Hellman
- querying properties of the capsule and of the cryptographic token.

Providing not only functions for creating electronic signatures, but also for verifying signatures relieves applications of the burden of dealing with this complicated subject area for the lifetime of the document. Applications are even ignorant to which algorithm (e.g. RSA, DSA, ECDSA) or which certificate format (e.g. X509, PGP, SPKI) is being used. Thus, not only can applications be light-weight, they also need not be updated when a new signature algorithm, a new signature format, or a new certificate format is introduced. This forward compatibility is one of the key benefits of the security-layer.

Access rights and their management are another area of critical importance. Again, the design relieves applications of dealing with this issue. Instead, applications issue their request through the interface ignorant of any access rights. The security capsule then either grants or denies access to the function according to its own security policy. This policy may involve the user, such as entering a PIN code, but it can also make automated decisions, e.g. based on the certificate of the application when using TLS or HTTPS as transport layer. Again, new technologies such as using biometric data instead of PIN codes can be introduced without the need to update applications.

It is important to note that the security-layer interface allows specifying Internet references (unique resource identifiers, URIs) instead of data itself making full use of the network wherever it makes sense. Thus it is not necessary that all data is transmitted through the security-layer interface. Instead the security-capsule resolves the references and downloads the data from the specified remote resource when desired.

4.1.1   Creating and Verifying CMS Signatures

The command for creating a CMS signature [10] takes a single file as input. The application can specify which key should be used for signing, and it also provides information on the file's MIME-type [21] [22], so that the capsule can invoke the appropriate trusted viewer for the file. The function returns a CMS object, which additionally contains information according to ETSICMS [11]. In particular, it contains a signed certificate reference. This is necessary, because otherwise one has no guarantee that the certificate used for signing is the one accompanying the CMS signature object.

CMS signatures can be created as detached signatures (data and signature in different files) or as enveloping signatures (CMS object encapsulating the data file).

The verify command takes a CMS signature object as parameter and optionally a data file in case of a detached signature. It returns two results:

one for the validity of the signature itself, and one for the validity of the signing certificate at a specified point in time.

4.1.2   Creating and Verifying XMLDSIG Signatures

The command for creating XMLDSIG signatures [20] is more sophisticated than its CMS counterpart, as the XMLDSIG recommendation has more features. It can take more than one data file as parameter. Each data file can be transformed by algorithms specified in XMLDSIG before the signing process. Optionally applications can provide additional supplements needed for these transformations.

Among the transformations allowed are XML canonicalization [23], XPath transformations [24], and XSLT stylesheet transformations [25]. The latter two transformations play an important role for usability and generalized data handling. The resulting XMLDSIG signature includes signed properties according to ETSI's "XML Advanced Electronic Signatures" [12], again in particular a reference to the signing certificate. Furthermore, the signature contains a manifest with references to all input data used for the transformations. This allows applications to make reasonable assumptions about the correlation between original input data and the transformed data actually signed.

The verify command takes a document containing an XML signature, an XPath describing the location of the signature inside the document, and optionally supplement data used for transformations. The function returns results for the validity of the signature, the validity of the signing manifest, and the validity of the signing certificate at a specified point in time.

4.1.3   Storing and Retrieving Data

Applications may store or read data from the security capsule. In some cases this data may be stored on the cryptographic token itself, in other cases this data may reside inside the capsule, or even somewhere on the net. The security-layer interface defines so called info-boxes, which are containers for data. Each info-box has an identifier (analogous to a file name) and contains data according to its type. There are two different types of info-boxes:
− a binary file type and
− an associative array type.

As the name already suggests, the binary file info-box behaves like a file. Applications can read the info-box, or overwrite the info-box with data. It is not possible to read or overwrite parts of the info-box, the command always affects the whole info-box content.

The associative array stores data in key/value pairs. E.g. the certificate info-box is an associative array, where certificates are stored in the pair-value and where the corresponding pair-key is set to the name used for selecting the certified signature key to be used when creating signatures. The interface has functions for creating, updating, and deleting pairs and for renaming and searching keys.

4.2   Transport Layer Binding

The security-layer protocol itself may utilize different transport layers. Transport layers currently defined are TCP/IP, TLS [26], HTTP [27], and HTTPS [28]. TCP/IP and TLS just transmit the XML requests and responses as they are. In case of TLS the capsule may evaluate the application's certificate when deciding whether to grant access privileges or not.

The HTTP and HTTPS bindings are designed so that Web browsers do not need any active components (not even scripts) to access the security-layer. To that end, the HTTP binding uses the standard mechanisms used by HTML forms. The binding defines a set of input fields:

– *XMLRequest:* this field holds the XML coded security-layer request itself.
– *DataUrl:* specifies an URL where the security-capsule should send the resulting response.
– *RedirectUrl:* specifies an URL to which the browser should be redirected in response to its request.
– *StylesheetUrl:* if no RedirectUrl is specified, the browser would directly receive the XML encoded response. In most cases however, one would rather not see the user receiving the XML as such. By specifying a XSLT compliant stylesheet the security-capsule uses this stylesheet to transform the XML response (for example into HTML) before sending it to the browser.

Among the fields listed above, only the first one (XMLRequest) is mandatory. By having the option to send the response directly to the server and send a formatted reply to the browser this binding offers utmost flexibility in design of the data flow and user experience.

CONCLUSIONS

The paper has given an overview to the approach followed by the Austrian e-Government initiatives. Two aspects have been identified as crucial in order to reach acceptance: On the one hand, security is a must to both achieving citizen's acceptance and to fulfill legal requirements. On the other hand, discriminatory situations ruling out solutions that may show up in the market in the future need to be avoided.

Both aspects have been fulfilled by means of an open interface – a so-called security-layer. The interface de-couples the application from the security-relevant functional blocks, such as the signature-creation process, by defining a general requirement catalogue. The market then can take up the approach by implementing the requirements with the technologies of choice. This results in the forward-compatibility aimed.

REFERENCES

[1] Austrian signature law: "Bundesgesetz über elektronische Signaturen (Signaturgesetz - SigG)", BGBl. I Nr. 190/1999, BGBl. I Nr. 137/2000, BGBl. I Nr. 32/2001.

[2] Austrian signature order: "Verordnung des Bundeskanzlers über elektronische Signaturen (Signaturverordnung - SigV)", StF: BGBl. II Nr. 30/2000.

[3] Administration reform law: "Verwaltungsreform Gesetz", 2001 amending "Allgemeines Verwaltungsverfahrensgesetz (AVG)" BGBl. Nr. 51/1991.

[4] Notification delivery law: "Bundesgesetz vom 1. April 1982 über die Zustellung behördlicher Schriftstücke", BGBl. I Nr. 137/2001.

[5] Posch R., Leitold H.: "Weissbuch Bürgerkarte", Bundesministerium für öffentliche Leistung und Sport, IT-Koordination des Bundes, June 2001.

[6] Directive 1999/93/EC of the European Parliament and of the Council of 13. December 1999 on a community framework for electronic signatures.

[7] European Electronic Signature Standardization Initiative: "EESSI explanatory document: Description of deliverables", EESSI Steering Group, 2000.

[8] International Organization for Standardization: "Information technology - Security techniques - Evaluation criteria for IT security", ISO/IEC 15408-1 to 15408-3, 1999.

[9] CEN/ISSS WS/E-Sign Workshop: "Security Requirements of Secure Signature Creation Devices (SSCD-PP)", CWA 14168 and CWA 14169, 2002.

[10] Hously, R.: "Cryptographic Message Syntax (CMS)", IETF Request For Comment RFC 2630, 1999.

[11] ETSI SEC: "Electronic Signature Formats, v.1.2.2", Technical Specification ETSI TS 101733, 2000.

[12] ETSI SEC: "XML Advanced Electronic Signatures (XAdES)", Technical Specification ETSI TS 101903, 2002.

[13] ETSI SEC: "Policy requirement for certification authorities issuing qualified certificates, v1.1.1", Technical Specification ETSI TS 101456, 2000.

[14] CEN/ISSS WS/E-Sign Workshop: "Cryptographic Module for CSP Signing Operations – Protection Profile (CMCSO-PP)" , CWA 14167-2, 2002

[15] European Electronic Signature Standardization Initiative: "Algorithms and Parameters for Secure Electronic Signatures, v2.1", EESSI algorithm group, 2001.

[16] RSA Laboratories: "RSA Cryptography Standard", PKCS #1 v2.1 draft 2, 2001.

[17] National Institute of Standards and Technology, "Digital Signature Standard (DSS)", NIST FIPS Publication 186-2, 2000.

[18] American National Standards Institute, "Public Key Cryptography for the Financial Services Industry: The Elliptic Curve Digital Signature Algorithm (ECDSA)", ANSI X9.62-1998, 1998.

[19] International Organization for Standardization, "Information technology - Security techniques - Cryptographic techniques based on elliptic curves - Part 2: Digital signatures", ISO/IEC FCD 15946-2, 1999.

[20] Eastlake D., Reagle J., and Solo D.: "XML-Signature Syntax and Processing", W3C Recommendation, 2002.

[21] Freed N., Borenstein N.: "Multipurpose Internet Mail Extensions (MIME) Part Two: Media Types", IETF Request For Comment RFC 2046, 1996.

[22] Murata M, Laurent S. St., and Kohn D.: "XML Media Types", IETF Request For Comment RFC 3023, 2001.

[23] Boyer J.: "Canonical XML", W3C Recommendation, 2001.

[24] Clark J., DeRose S.: "XML Path Language", W3C Recommendation, 1999.

[25] Clark J.: "XSL Transformations (XSLT)", W3C Recommendation, 1999.

[26] Dierks T., Allen C.: "The Transport Layer Security (TLS) Protocol, Version 1.0", IETF Request For Comment RFC 2246, 1999.

[27] Gettys, Mogul, Frystyk, Masinter, Leach, and Berners-Lee: "Hypertext Transfer Protocol, HTTP/1.1", IETF Request For Comment RFC 2616, 1999.

[28] Rescorla: "HTTP over TLS", IETF Request For Comment RFC 2818, 2000.

CADENUS SECURITY CONSIDERATIONS

Gašper Lavrenčič[1], Borka Jerman-Blažič[2], Aleksej Jerman Blažič[3]
[1]SETCCE, [2]University of Ljubljana, [3]Institut Jožef Stefan, Ljubljana, Slovenia

Abstract: This paper deals with the development of security service provision required for end-user services in Premium IP networks that are being developed and tested within CADENUS project funded under an EU initiative. The initiative is focused towards setting up and experimenting with mechanisms that enable dynamic user-services creation, configuration and provisioning with in-built QoS. The authentication, authorization, data integrity and non-repudiation services are provided through exchange of information and documents between the mediators communicating with end user, the network resource manager and the service manager acting in the middleware part of a Premium IP network. The exchange is using the technology specified in the Extensible Markup Language (XML) messaging with the additions taken from the emerging Electronic Business XML (ebXML) model. The paper presents the XML/ebXML specified security mechanisms, which are in-built in the CADENUS architecture.

Key words: XML, ebXML, digital signature, Access Mediator, Service Mediator, Resource Mediator, Collaboration Protocol Profile/Agreement, Messaging Services

1.   INTRODUCTION

The current Internet is based on the best-effort model, which does not provide any traffic segregation or differentiation within the network. Guaranteed delivery or high-priority delivery provides a much-needed alternative to the best-effort networks of today. In order for networks to continue to grow, to satisfy new service needs, and to expand to serve real-time applications, networks must provide mechanisms that ensure delivery within sensible bounds or offer preferential treatment to certain traffic. In recent years, much industrial and academic effort has been devoted to the definition of schemes and architectures that provide guarantees to data communication carried over packet-switched networks. These efforts have achieved several interesting results: the Integrated Services (IntServ) model [10], the Differentiated Services (DiffServ) model [11], Multi-Protocol Label Switching (MPLS) [12], etc. by the IETF [13] community, as well as proposals for the efficient integration of IP and ATM are among the most prominent examples. However, in the case of the IETF, much of the work is yet to be completed in terms of real specifications and implementation in running networks. In this document, Premium IP networks refer to these new IETF network technologies known as the next generation of IP networks.

The project "Configuration and Provisioning of End-User Services in Premium IP networks" or in short CADENUS is an R&D project from the 5th Framework Program funded by the European Commission. The main focus of the project is integration of existing research in Premium IP networks and end-user services with QoS guarantees. Current DiffServ developments within the IETF have presented an architecture and framework to enable Premium IP network services to be configured and provisioned. These proposals do not, however, describe concrete realisations for networks providing Premium IP using Differentiated Services. In addition, the Premium IP packet stream differentiation is the only aspect of the end-user service offering. Many other problems that deal with the provision of these services on network level are not sufficiently elaborated. In this context a service provider dealing with traffic streams that require multiple QoS traffic streams has to address the bearer level service as well as higher-level service alike. The CADENUS project proposes an integrated solution for the configuration and provisioning of end-user services with QoS guarantees in Premium IP and is exploring the possibility and benefits of bringing classes of end-user services into a unified framework.

The CADENUS framework is a set of core functionality at the user - provider interface in Premium IP. The service creation and configuration is provided in a dynamic way by linking user-related service components (authentication, authorization, registration, subscription, etc.) to network related service components (QoS control, resources management, accounting, call control, etc.). This paper discusses the security solutions proposed and implemented within CADENUS framework that links relevant end-user request and network related service components.

2. SHORT OVERVIEW OF THE CADENUS ARCHITECTURE

The architecture of the integrated solution for provision of end-user services in Premium IP networks is based on elements that support service creation and service configuration in QoS aware IP networks. Closely related to this activity is the management of those resources on the underlying networks that are reserved on registration/subscription, and those that are used and maybe subsequently modified – when the service is invoked/configured. Associated with the reservation and usage of resources is the automated production and presentation of the corresponding SLAs (Service Level Agreement) to the user, and SLA-SLS (Service Level Specification) translation. CADENUS architecture (see Figure 1) introduces 3 components that are necessary to supervise the dynamic service creation configuration process: Access Mediator (AM), Service Mediator (SM) and Resource Mediator (RM).

The architectural platform is distributed among various mediators for its flexibility and guaranty of QoS. Each mediator is designed for its administration environment to ensure transparency of the services offered (e.g. backbone providers, internet service providers).

AM is a device into which users input their requests to the system. Whilst optional, the AM adds value for the user, in terms of presenting a wider selection of services, ensuring the low service cost, and offering harmonised interface. In the CADENUS model AM presents the current service offer to the user. The source of services is stored in the Service Directory database (SDDB). The AM maintains some of the elements allowing access control for the end user in order to assist and ease the service selection process. The usage of the service involves two business processes: registration of the user to the service and invocation of the service at the moment it is used (change of parameters of the same services during a session is considered as new invocation). AM may form associations with one or more Service Mediators to which requests are issued. SM is responsible for finding and in some cases, building from individual elements – the requested service, through a process that involves information request from the Resource Mediators and later selection of the appropriate Resource Mediator. Resource Mediators are responsible for selection of the appropriate network capabilities from the available options in the underlying network. The service creation and service configuration is achieved from components of all architectural elements (AM, SM, RM) and is presented to the user in the form of SLA.

The Service Mediator (SM) supervises the incorporation of new services in .the SDDB and the management of the physical access to these services via the underlying network, using the Resource Mediator(s) - RM. AM prepares a SLA, containing data about a service chosen by a user (e.g VoD, Best Quality), passes it to SM, that transforms data from SLA to service technical specification document –SLS, (e.g Best Quality=2 Mbit/s) and passes it to the RM.

The SM has an important role as this is the place where created services are attached, and from where the impacts of service reconfigurations are communicated to the network resource management. A task of the SM is also to make information about new service offerings available to AM. SM is also responsible for refreshing and up dating the system responsible for management of all services with new rules (if they appear) and configurations of the devices affecting the network resource management functionality.

Reservation of the network capacities is done with communication between the SM and RM. This communication is generic and is independent of the used network technology. The reservation of the network capacities is specified in the SLSs. In the CADENUS architecture the RM is responsible for the end-to-end view of the network QoS. This is done through communication of the RM with the appropriate underlying network management systems. A network provider wishing to offer its resources must support an interface capable of handling an SLS, from its network management system to one or more RMs. In order for the RM to maintain and up date the end-to-end network view of the current QoS availability, it may use a set of policy that is agreed with the underlying network management systems.

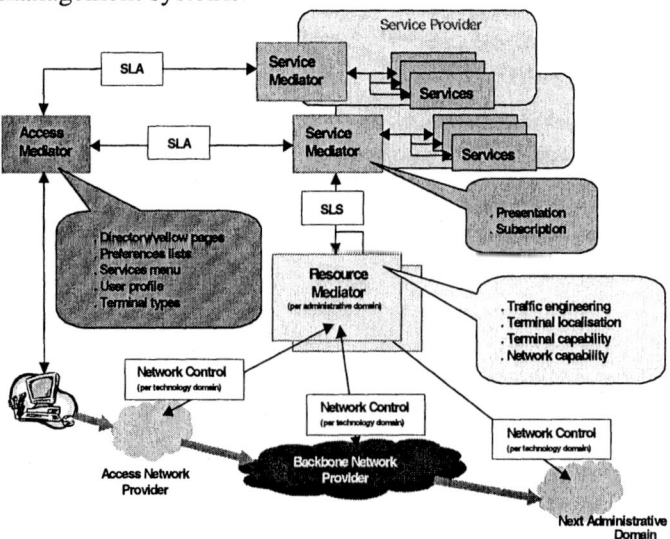

Figure 1. Basic CADENUS architecture

Communication and coordination within the architectural platform is based on XML messaging technique. Extendable Markup Language was chosen to provide transparency and interoperability of the overall architecture. Requests, demands and responses are processed using dynamic XML document generation. Due to the specific request regarding the security provision and the constrains that appeared in the deployment of the platform several new mechanisms have been proposed. Electronic business XML specification was chosen for production of the specification of the overall CADENUS architecture.

3. SECURITY CONSIDERATIONS

The security considerations in the CADENUS architectural model include all aspects of communications e.g. between a client and the platform, between the elements of the platform and between the platform and the underlying network.

Access Mediator acts as an interface between a client and service providers. AM, for example, can select the cheapest offer of a movie available within the offers of more service providers, it can notify the user immediately when a new movie becomes available that matches the stored user profile. In accessing the platform through AM *authentication* is required on the user request level. When authenticated, the user requirements are then captured, and the AM can send the acquired information to the SM who then employs the RM(s) to map the request and to provide the selected service into the physical network. A Trusted Third Party that is a part of a value-added service provider can be involved in authentication. A task of the AM is also to register the users and keep information about their access control parameters.

After the service selection has been agreed with all parties, then a SLA needs to be "signed" between the AM and the SM(s) – in order to provide *data integrity* and *non-repudiation* of the requested and offered service. AM is responsible for the preparation of the SLA. The SM then transforms the SLA into a SLS and passes it to RM(s) enabling the service to be put in place. The provision of integrity and non-repudiation is also provided with digital signature on the SLS. The communication between AM and SDDB also requires data integrity protection as well as authentication of the communicating parties. Here, again the Trusted Third Party services for key management purposes are used.

The RM is associated with the underlying network and its capabilities that provide QoS, but the communication between the SM and the RM is generic (i.e. independent of the technology employed by the underlying network). RM collects the technical data (e.g. the necessary bandwidth for provision of user-selected service) related to the services from the SM via SLSs, and passes them to the network dependent Network Controller (NC). NC is an entity that contains the underlying DiffServ and/or MPLS network configuration architecture data.

3.1   Selected technology

The CADENUS platform is built on mechanisms for process co-ordination and as a consequence the ebXML specification seemed to be natural tool for specification of these processes and their interactions.

EbXML has its roots in the requirements of business environment for enterprises that dynamically conduct business with each other in an open market environment. The major tasks specified within XML technology are:
-   Enabling discovering of products and services in an open on-line market offer.
-   Enabling shared business processes and associated document exchanges.
-   Setting up contact points and exchange of information for relevant businesses.
-   Enable contractual terms for chosen business processes and associated information.

EbXML is being designed to meet the needs for information exchange in business environment in an automated way and as such ensures three basic concepts to be met:
-   Infrastructure that ensures data communication interoperability,
-   Semantics framework that ensures commercial interoperability and
-   Mechanisms that allow enterprises to find each other in the open market on line, agree to become trading partners and conduct business with each other.

The following features of the ebXML specification were relevant in the CADENUS specifications relevant to the security considerations:
-   Messaging services.
-   Registry and Repository system.
-   Collaborative Partner Agreements.

The ebXML Messaging Service [4] specification defines a set of services and protocols that enable electronic applications to exchange data via ebXML messages in the Messaging service definition part of ebXML. The specification allows well-established cryptographic techniques to be used for implementation of strong security mechanisms in the message exchanging process, e.g. HTTP over Secure Socket Layer (HTTPS) can be used for provision of confidentiality, data integrity and authenticity, as well as digital signatures can be applied to individual messages for provision of authenticity, message integrity authorisation and non-repudiation.

The Registry & Repository part of ebXML specification provides storage rules for all published ebXML documents. It enables storage of company/user profiles and trading partner specifications. The Registry & Repository provides many key functions. For the user it stores company profiles and trading partner specifications. It provides mechanisms for the trading partners to find each other's company business profile.

The Collaborative Partner Agreement specification [3] defines the technical parameters of the Collaborative Protocol Profile (CPP) and the Collaboration Protocol Agreements (CPA). CPP is an XML document that contains technical specifications of the communicating partner and its roles in the business processes. The CPA is an XML document representing technical interception of two (or more) CPPs.

The CADENUS defined SDDB is implemented as a Registry & Repository where the business process specification documents for each type of the service (VPN, VoD, VoIP) are stored. The Service Mediator that is in the Service Provider's domain creates different business process specification documents [2], e.g. one for each service type (VoD, VPN or VoIP), according to the ebXML Process Specification Schema. According to the created documents the Service Mediator creates the Collaboration Protocol Profile ebXML document(s) defining its collaboration role in business process, and stores it in the ebXML Registry & Repository. The Access Mediator creates (in the same way the Service Mediator does) Collaboration Protocol Profile ebXML document according to the Business Process Specification Document(s) in ebXML Registry & Repository. At the service invocation time a user claims his service requirements to the Access Mediator. The Access Mediator browses the Registry & Repository and downloads the Service Mediator Collaboration Protocol Profile, and makes user-understandable information about the service. If a user agrees on the terms the Collaboration Protocol Agreement ebXML document representing the contents intersection of both CPPs is created and installed in the Access Mediator. A communication channel for ebXML messages (SLAs) is established between the Access Mediator and the Service Mediator. The Service Mediator defines the technical data, packs them into SLSs, and passes to the Resource Mediator. The message exchange between the Service Mediator and the Resource Mediator is via XML documents, but itself doesn't belong to the ebXML infrastructure.

3.2   Implementation

ebXML messages used for communications over the Internet are exposed to several third party attacks. From security point of view the most exposed parts of the ebXML architecture are messages exchanged between an AM and the Registry & Repository (CPP, optionally CPA), SLAs exchanged between AM and SM, as well as SLSs between SM and RM.

The disclosure of these messages is not very critical and the confidentiality provision is not necessary. The security concern here is to disable unauthorized parties to alter the message contents and thus message integrity and authentication must be provided as well as reliable delivery. ebXML Messaging Service provides elements for reliable message delivery via implementation of the protocol for message sending repetition on the sender's side. When a message arrives to the recipient an acknowledgement message is created and sent back to the party whose message is acknowledged. In addition to that, the message is digitally signed, providing non-repudiation.

The data integrity of the Service Mediator CPP(s), the Access Mediator CPP and the created CPA within Access Mediator referencing to the business process specification document(s) within the ebXML Registry & Repository is provided by use of XML signatures [6]. This is done in **ProcessSpecification** element described in CPPs and a CPA.

The signature generation steps follow the W3C/IETF recommendations:

3.3   Create a **ds:SignedInfo** element with **ds:SignatureMethod**, **ds:CanonicalizationMethod** and **ds:Reference** elements for the SOAP Header and any required payload objects.

3.4   **Canonicalize** and then calculate the **ds:SignatureValue** over **ds:SignedInfo** based on algorithms specified in **ds:SignedInfo**.

3.5   Construct the **ds:Signature** element that includes the **ds:SignedInfo, ds:KeyInfo**, and **ds:SignatureValue** elements.

Example of a digitally signed ebXML SOAP Message can be seen on Figure 2

```
<?xml version="1.0" encoding="utf-8"?>
<SOAP-ENV:Envelope
  xmlns:SOAP-ENV="http://schemas.xmlsoap.org/soap/envelope/"
  xmlns:eb="http://www.ebxml.org/namespaces/messageHeader"
  xmlns:xlink="http://www.w3.org/1999/xlink">
  <SOAP-ENV:Header>
    <eb:MessageHeader eb:id="…" eb:version="1.0">
    ...
    </eb:MessageHeader>
    <eb:TraceHeaderList eb:id="…" eb:version="1.0">
      <eb:TraceHeader>
      ...
      </eb:TraceHeader>
    </eb:TraceHeaderList>
    <ds:Signature xmlns:ds="http://www.w3.org/2000/09/xmldsig#">
      <ds:SignedInfo>
        <ds:CanonicalizationMethod
Algorithm="http://www.w3.org/TR/2000/CR-xml-c14n-20001026"/>
        <ds:SignatureMethod
Algorithm="http://www.w3.org/2000/09/xmldsig#dsa-sha1"/>
        <ds:Reference URI="">
            <Transforms>
                <Transform Algorithm="http://www.w3.org/TR/1999/REC-
xpath-19991116">
                  <XPath
xmlns:dsig="http://www.w3.org/2000/09/xmldsig#">
                      not(ancestor-or-self::eb:TraceHeaderList or
                      ancestor-or-self::eb:Via)
                  </XPath>
                </Transform>
            </Transforms>
        <ds:DigestMethod
Algorithm="http://www.w3.org/2000/09/xmldsig#dsa-sha1"/>
        <ds:DigestValue>...</ds:DigestValue>
        </ds:Reference>
        <ds:Reference URI="cid://blahblahblah/">
        <ds:DigestMethod
Algorithm="http://www.w3.org/2000/09/xmldsig#dsa-sha1"/>
        <ds:DigestValue>...</ds:DigestValue>
        </ds:Reference>
      </ds:SignedInfo>
      <ds:SignatureValue>...</ds:SignatureValue>
      <ds:KeyInfo>...</ds:KeyInfo>
    </ds:Signature>
  </SOAP-ENV:Header>

  <SOAP-ENV:Body>
    <eb:Manifest eb:id="Mani01" eb:version="1.0">
      <eb:Reference xlink:href="cid://blahblahblah"
        xlink:role="http://ebxml.org/gci/invoice">
        <eb:Schema eb:version="1.0"
eb:location="http://ebxml.org/gci/busdocs/invoice.dtd"/>
      </eb:Reference>
    </eb:Manifest>
  </SOAP-ENV:Body>
```

Figure 2. Digitally signed ebXML SOAP Message

The key authentication is to be provided by CAs (Certification Authorities). Additionally, XML Key Management Services, specified by W3C, can be used to delegate some key management functions to a Trusted Service and thus reduce the complexity on the entity's side. The decision which solution will be used depends, on the processing capabilities of the Mediators on one hand, and the Trusted service's server on the other, and the composition of ebXML messages itself. For example, several signatures in a document may use a key verified in an X.509 certificate chain appearing remotely; each signature's **ds:KeyInfo** element can reference this chain using a single Uniform Resource Locator (URL), instead of including the entire chain of X509Certificate elements. The entity that is verifying the signature can then retrieve the whole chain and validate the key, or delegate this task to its Trusted Service.

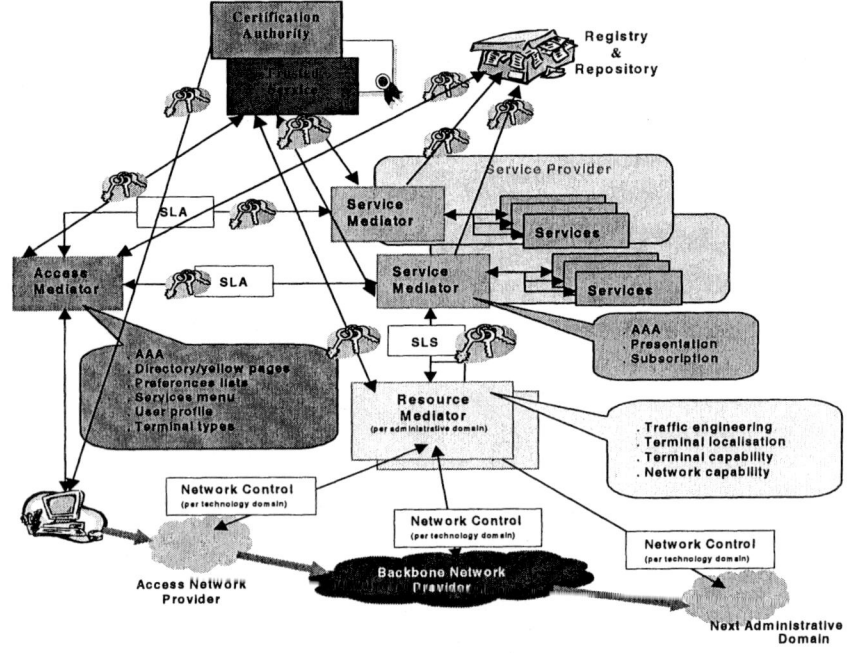

Figure 3. CADENUS secure ebXML/XML architecture

Figure 3 presents the CADEUNS ebXML/XML secure architecture, where keys present secure communication links for message exchange.

VALIDATION OF LONG-TERM SIGNATURES
About Revocation Checking of Certificates in the Context of long-term Signatures

Karl Scheibelhofer
Institute for Applied Information Processing and Communications
Graz University of Technology
Inffeldgasse 16a
A-8010 Graz
Email: Karl.Scheibelhofer@iaik.at

Abstract: The current practice of digital signature creation is simple. However, signature verification is much harder. This especially holds for long-term signatures, signatures that should remain verifiable over years. After the signing certificate expired, it is hard to find out if the certificate was valid at the time the signature was created. Current revocation checking mechanisms, like CRLs and OCSP, may not provide the status of certificates which are no longer valid. This is one reason why many of the current signature verification systems cannot verify signatures after the signing certificate expired. There are several approaches for coping with these problems: attach all data that is required for validation to the signature right after signature creation, let the verification software collect and archive all validation data that it needs, or use advanced services for certificate status checking. Currently, there are hardly any advanced services available. However, this paper shows that it is not hard to design such services.

Key words: digital signatures, long-term signatures, revocation checking, certificate status checking, advanced electronic signatures, CRL, OCSP, DPV

1. INTRODUCTION

Not only since the European Union published the Directive on Electronic Signatures [1] digital signatures have been a big issue. Many European countries have enacted laws to implement this directive [2]. There are already CAs (Certification Authorities) which provide solutions that are compliant with the laws. Such products often use a smart card as signature creation device, a smart card reader, software for signing and verification, and a qualified certificate.

Everyone who works with signed email knows how easy it is to sign a document. Insert the smart card into the reader, enter the PIN (Personal Identification Number) and then the smart card calculates the signature value over the data to be signed. For the process of signing, most systems do not need to access any online service. If the user created the signature using products compliant with law, the resulting signature will be a legally valid electronic signature. However, this is just half of the story. Having a legally valid signature is nice, but we also need to be able to verify the signature.

Verifying the signature right after it was created can also be harder than one might think. This is the case with systems that use CRLs (Certificate Revocation Lists) [5]. Because CAs are not required to issue a new CRL each time a certificate gets revoked, the client will not notice a certificate's revocation before the CA issues the next CRL. From the most recent CRL the client can only get reliable status information for times before the CRL was issued. This problem also occurs with OCSP (Online Certificate Status Protocol) [6] responders which are implemented based on CRLs. Unfortunately this is very often the case.

When the CA issued the next CRL after the signature creation, CRLs and OCSP responders will provide the necessary information to verify the signer's certificate. To verify a signature, it is required to verify the certificate with respect to the time when the signature was created. Current mechanisms may not provide this information directly. This makes the verification of signatures hard, especially if one needs to verify them after the certificate expired. In this case, current revocation checking mechanisms may not provide any information about the certificate. Because of this, current software products that use digital signatures are unable to verify signatures after some time. For instance, most of the prominent email clients have this problem, but most of all other software products that verify signatures have the same problem. However, there are ways to cope with this problem. This paper will show some of them. One approach tries to solve this problem at creation time of the signature. It attaches all validation data that the verification process needs to verify the signature. Another approach requires new revocation checking mechanisms. Such new mechanisms can ease signature validation dramatically. This paper proposes a simple but effective revocation checking service.

2. VERIFICATION OF LONG-TERM SIGNATURES

It is easy to sign a document using current technology. All you need is signing software, a signature key, and a certificate for this key. You can get all these things easily. When a user signs a document he uses his signing key to create a signature value. The signed version of the document will consist of several parts: the original document, a signature value and the signer's certificate (maybe not the complete certificate, but at least an unambiguous identifier of it). Moreover, the signed document may contain other attributes like the signing time or an identifier of the policy under which the user signed the document. Figure 1 shows the contents of such a signature. The signature value covers all those elements which are highlighted grey. But is all this information enough to be able to verify the signature later on?

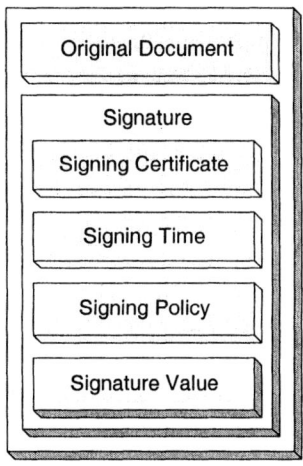

Figure 1. The contents of a simple signed document.

We can find this out by having a closer look at software that tries to verify the signature. Let us assume that a relying person tries to verify the signed document one year later. In addition, we assume that the certificate that the signer used is expired meanwhile. The software can use the certificate, which is contained in the signed document, to cryptographically verify the signature value, even though, this is not enough. To verify the signature, the software must also find out, if the signer's certificate was valid at the time the signature was created. If the certificate is valid now, at the time of verification, is irrelevant. The signing time is normally part of the signed document. Maybe, the user can specify the signing time when verifying the signature, if the signed document does not contain the signing time. Having the signing time, it must find out, if the certificate was valid at this time. The following sections show different methods for that.

2.1　Using CRLs for Revocation Checking

The software may try to use CRLs to get the desired information. Almost all CAs provide CRLs for their issued certificates. In principle, CRLs are lists of certificates which have been revoked. The list does not contain the complete certificates but the serial numbers of the certificates. The issuer of the CRL is also given in the CRL. The combination of issuer and serial number identifies a certificate uniquely in X.509 based systems. Thus, the software checks if the serial number of the concerned certificate is in the CRL. If the certificate is within its validity period and the serial number is not in the CRL, the certificate has not been revoked and can be considered valid. A CA issues CRLs regularly. Each CRL contains the date when it was issued and the date when the next CRL will be issued latest. With this information, the client can see, if the CRL he has is still valid. The CA may issue a new CRL even earlier than the date given in the last CRL. This may be applicable, if some certificate has been revoked meanwhile and the CA wants to provide current revocation information. Thus, a client may decide to try to get always the latest CRL. Figure 2 shows the most important elements of a CRL. The extensions allow conveying additional information, for example, an identifier of the signing key of the authority or the number of the CRL, which is a monotonically increasing sequence number.

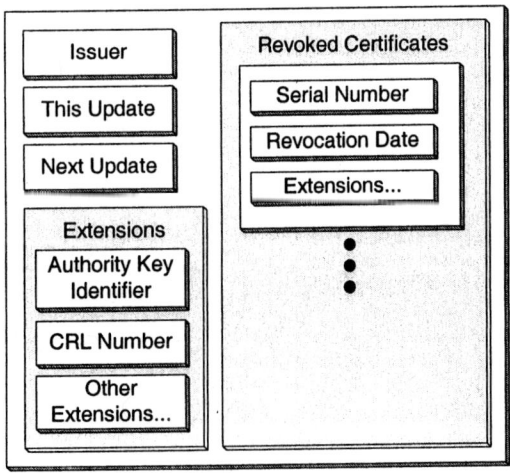

Figure 2. The most important contents of a CRL.

To find out where to get the CRL, the software can investigate the certificate itself. Inside a certificate, there is an entry that specifies where the software can find the CRL for this certificate - this is called the CRL distribution point. In practice, this is often an URL (Uniform Resource Locator) to a file on an HTTP (Hypertext Transfer Protocol) sever which contains the most recent CRL. There we can already see a problem. Under this location, the client only gets the current CRL. The current CRL normally contains only information about certificates which are still within their validity period. However, we need a CRL, which contains information about the status of the certificate at the signing time – for instance, one year ago. There is no standard mechanism to get old CRLs. So what can the software do in this case? Instruct the user to contact the CA for getting an old CRL that was valid at the signing time, download it manually and configure the software to use this? – Most users will say something like that: "What the hell is a CRL? And why is my software not able to do that for me automatically?" And they are right. This is unsuitable for practical purposes. The only chance for the software to verify the certificate is to have all CRLs available locally - all CRLs of all CAs that the user may ever try to verify a certificate for. This is practically infeasible.

Do delta CRLs relieve the situation? Delta CRLs are a special form of CRLs. They aim at keeping the amount of transferred data small. A delta CRL just contains the changes with respect to a complete CRL. To be useful for the client, the client needs this base CRL and the delta CRL. The base CRL may contain hundreds of revoked certificates. For instance, after this base CRL has been issued, but before the next complete CRL gets issued, the CA revokes another two certificates. To avoid issuing a new complete CRL, the CA can issue a delta CRL which refers to the base CRL and contains the two additional certificates. Consequently, a client, which already has the base CRL, can download the latest delta CRL. This delta CRL will be relatively small compared to the complete CRL. Having the base and the delta CRL, the client has all the latest revocation information. The CA may issue delta CRLs even more frequently than CRLs. However, a delta CRL must always refer to a complete CRL; it can never refer to another delta CRL as its base. Similarly to complete CRLs, the certificate itself contains the information where the client can find a delta CRL. Now we can see, that delta CRLs do not provide any information that we cannot get from the complete CRL. Thus, we have not made a real progress with respect to our verification problem.

There is another type of CRL called indirect CRL. Indirect in this context means that the issuer of the CRL is not the same as the CA that issued the certificates that are in scope of this CRL. A CA can use indirect CRLs to outsource CRL issuing. In this case, the CA will issue a certificate for the other organization. This organization uses this certificate to issue and sign CRLs for the CA. Another situation where indirect CRLs are necessary is when a CA ceases operation. Then a different organization may continue to provide CRLs for this CA. In this case, the other organization will also use a different key for signing the CRL – it will be an indirect CRL. All in all, there is not much difference between normal CRLs and indirect CRLs, at least not regarding the information they provide for the client.

2.2   Using OCSP for Revocation Checking

Let us assume that we have a more sophisticated software. A software that is able to use OCSP services. OCSP is an online service that normally uses HTTP as transport protocol. If the certificate contains the address of the OCSP responder, the software can use it. The client sends a request to the OCSP server and gets a response. With such a request, the client can ask for the status of a certain certificate. The answer of the server tells the client one of three possible states: the certificate has been revoked with the revocation time, the status is good or the status is unknown. If the certificate has been revoked, the meaning of the answer is clear. The meaning of unknown status is also clear, though not very helpful. What does "good" mean? It does not mean that the certificate is valid. OCSP defines "good" as that the certificate has not been revoked. However, this does not imply that the certificate is valid, is within its validity period or has ever been issued.

Now, all is fine, the software will be able to verify the certificate, won't it? Unfortunately no. OCSP, as it is now, may not provide any additional information to CRLs. Depending on the implementation, an OCSP responder may not provide any more information than the most recent CRL. But doesn't it tell me the status of certificate at the signing time? No, not directly. OCSP does only provide the current certificate status. Seen realistically, OCSP is no big advance over CRLs. However, it is possible to implement an OCSP responder in a manner that it provides more information than CRLs do. It can provide the current status of a certificate; this is the status of the certificate when the OCSP server creates the response. Moreover, it can provide the certificate's status as long as it is required, long after the certificate expired. Many current implementations do not provide any more information than the recent CRLs, because the OCSP standard does not require it. Using such an implementation, OCSP only saves bandwidth in some cases. As we can see, software that is based on current revocation checking mechanisms may be unable to check the certificate's status in the depicted use-case – validation of long-term signatures without having the necessary validation data in the signed document.

2.3 Adding Validation Data to the Signature

The only solution that can be implemented using the current standards is adding all necessary data to the signature at signing time (or short after that). This means, attaching the complete certificate chain and all corresponding CRLs to the signed document. Remind that we need the first CRLs that were issued right after signature creation. The certificate chain may require all certificates up to a self-signed CA certificate, but in certain scenarios a shorter certificate chain may be sufficient. In general, the signer cannot anticipate in advance which CA certificate in the chain a verifier trusts. Additionally, the signing software may add a trusted timestamp to the signed document. Such a timestamp can provide evidence that the signature was created before the indicated time. This will support the verification software in determining the signature time. Instead of or in addition to CRLs, the signing software could also attach current OCSP responses. Figure 3 shows an example for such a signed document with attached validation data.

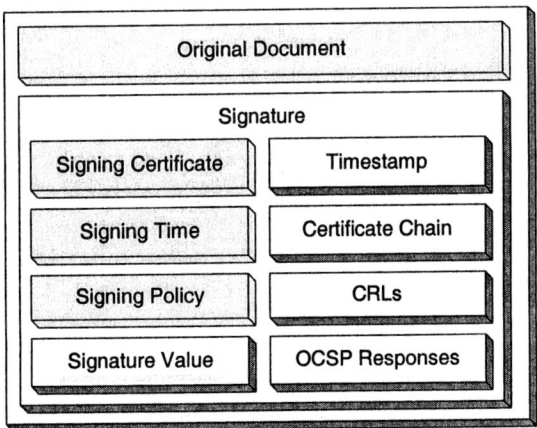

Figure 3. A signed document with attached validation data.

Adding the CRLs or OCSP responses would result in a considerable amount of extra data. Normally, the verifier needs to validate a complete certificate chain, which means that he needs the CRLs of all involved certificates. Moreover, if there is any indirect CRL involved, the verifier must construct and validate a separate certificate chain for the signer of this CRL. Attaching several CRLs and/or OCSP responses to each signature is not very convenient. In some use-cases, the required amount of additional memory for this validation data may exceed the original document's size. There are even standards from the ETSI that specify how to encode such types of electronic signatures, signatures that include all the validation data.

The ETSI TS 101 733 [3] document defines such structures for encoding signatures based on CMS (Cryptographic Message Syntax) [8] and the ETSI TS 101 903 [4] document defines similar structures for signatures based on the XML Signature [9] standard. Both standards are rarely used in practice yet.

2.4   Advanced Certificate Status Checking Services

What would solve such problems is a service that can answer the simple question "Was this certificate valid at this specific time?" Currently, there is no such service and it is unlikely that we will see such services in the near future. The PKIX (Public-Key Infrastructure X.509) working group is currently working on successor of OCSP versions one. This successor may include a service called Delegated Path Validation (DPV). This service could validate a complete certificate chain with respect to a certain time. The service will use the chain validation algorithm as specified by the PKIX working group. Such a service will come much closer to what we need to verify a signature, or more precisely to verify the certificate that was used for signing. The service must have access to the complete history of all certificates in its scope; otherwise it would not be able to answer our request.

This service would have to tell if the certificate chain, which ends with the signer's certificate, was valid one year ago, at the time the signature was created. If the service has only access to the current CRL, it would not be able to answer this request either. In this case, the service would respond that it does not know if the certificate chain was valid at the concerned time. We see, it is not sufficient to define such new protocols. The providers of such services must have databases that contain the required history of the certificate states, or they must have access to a service that provides the same information. If these services are implemented solely based on mechanisms like CRLs, they will not provide any additional information. Moreover, the service providers need to maintain these databases for many years; at least as long as the signatures should be verifiable. If a signature should be verifiable for at least thirty years, which may apply to advanced electronic signatures, the service provider needs to maintain the certificate's status history at least thirty years after the certificate expired. Seen realistically, this is not an extraordinary requirement. The laws for electronic signatures often require archiving the revocation information for such long periods. In addition, services that provide delegated path construction and validation may be suitable for enterprises but not for CAs. A CA may not want to provide such service or it may want to outsource it. In this case, the service providers must have means to get current information about certificate states. Services like delegated path validation will not solve the problem of getting the status of a certificate, they only move the problem to the service providers.

A problem with advanced services like DPV is trust. The client has to decide, if he trusts in the correct operation of that service. A user will not delegate such a relatively critical task to a service that he does not trust. Therefore, it is likely that each enterprise runs its own service, because trust relations are easier to manage in a closed environment. A public service will return a signed response. In consequence, the client has to verify the signature of the response. This, once again, requires verification of the certificate. The responsibility can never be transferred completely to a service; the final trust decision is always up to the client. In practice that means that the client needs a trust anchor. A trust anchor is most likely a certificate which the client trusts, normally a CA certificate. Careful users check the hash of trusted certificates, before they configure their client to trust this certificate. To check the hash, they use a trusted channel, which may be via telephone or even face to face.

2.5   Signed Document Store

Another solution for this problem would be a dedicated server which accepts signed documents and takes care about the collection of all necessary validation data. This approach would have several advantages over the straight forward one, which attaches all this validation data to the signed document. The server would collect each CRL just once and would store it locally; it can use the same CRL for other signed documents as well. Contrary, for the all-in-the-document approach, each client that signs a document collects the validation data at least once. Moreover, a server could parse incoming CRLs and convert the status information into a more efficient form. This may be a database that keeps track of the history of all concerned certificates. Such a server could also easily keep track of resigning. Resigning of signed document is required, if algorithms or keys become relatively weaker over time, because technology and science advance and it becomes easier to break the keys or the algorithms. Therefore, the server could resign the signed document (more precisely the signature of the document) with a stronger algorithm or key before the signature can be forged. If a signature has been resigned early enough with a stronger algorithm or key, it does not matter, if the inner signature is forgeable afterwards. Of course, after resigning we must obtain a timestamp that provides evidence that the signature was resigned before it could have been forged. Confidentiality of documents would not raise a problem, because it would not be required to send the original document to the server. The server does not need it to maintain the signature. The original document is only required for signature verification. Moreover, the original document could be sent encrypted.

2.6   Proposal for a Simple Certificate Status Checking Service

We saw that the current mechanisms for certificate status checking, CRLs and OCSP, are not optimal. However, a simple service could provide all information we need. First, we need to find out what information such a service should provide. The question that the client needs an answer for is "Was this certificate valid at this time?" Thus, the minimal information that the client must send to the service is: the certificate (or an unambiguous identifier of it) and the concerned time. The answer could be one of three: the certificate was valid at this time, the certificate was invalid at this time or the service does not know the answer. If the certificate was invalid or the service does not know the answer, the service may provide additional information about the reason. Such a service is relatively simple. Figure 4 shows the contents of possible requests and responses. The figure does not show it, but the response will contain a signature and the certificate of the responder. Based on the responder's certificate, the client can make its trust decisions. The trust model can be the same as for other certificates.

Upon setup of a new responder, we can get all status information about the past from CRLs. Of course, we would need all CRLs that a CA issued. In practice, the service would convert the information contained in the CRLs into a more convenient form; for instance, it would store the information in a database. As already mentioned, there is no automatic mechanism for getting old CRLs. Thus, the administrator needs to do this manually. This is not a very big drawback, because it is needed just once at setup time. After setup, the service gets the states directly from the database. The CA has to ensure that its database always contains the current states.

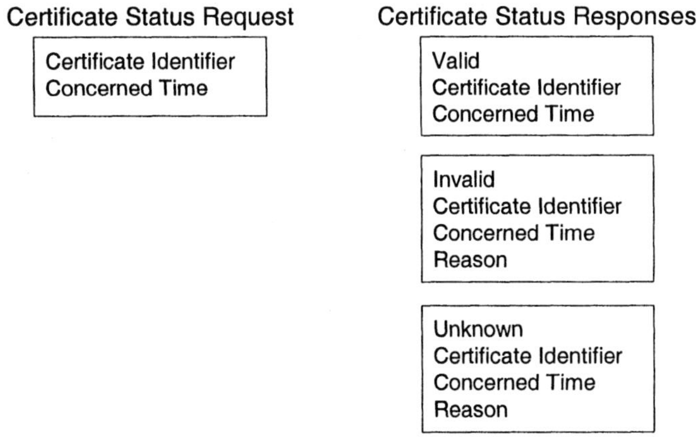

Certificate Status Request

> Certificate Identifier
> Concerned Time

Certificate Status Responses

> Valid
> Certificate Identifier
> Concerned Time

> Invalid
> Certificate Identifier
> Concerned Time
> Reason

> Unknown
> Certificate Identifier
> Concerned Time
> Reason

Figure 4. The contents of requests and responses of a simple certificate status service.

It is easy to extend this service to provide complete status histories of certificates. The only thing that we need to change in the request is the time field. We would provide a time interval instead of a single time value. The answer that this request would express would then change to: "What was the status history of the given certificate in the given time interval?" The response is a list of status information records. The first entry in the list is the status at the beginning of the time interval. There would be one additional entry in this list for each status change. Each of these entries would have a time value, which tells, when the status changed to the new state. In Figure 5 we can see the structure of such requests and responses. If we just want to know "Was this certificate valid at this time?", it is still easy to use. In such a case, the client sends a time interval of zero length; it starts at the concerned time and ends at the concerned time. The response of the server can only contain one entry in the list, because a certificate can only have one state at a single point in time. As in the simpler version, the responses will be signed and will contain the responder's certificate.

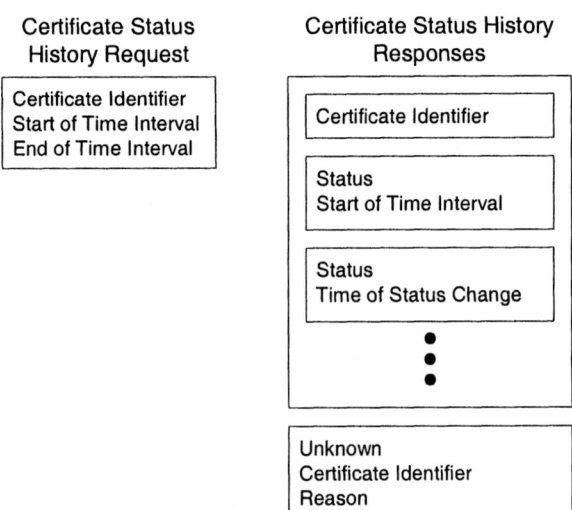

Figure 5. The contents of requests and responses of an advanced certificate status history service.

What advantages would this advanced version have? A client can get the complete history of a certificate with one request/response. For example, if a client has a certificate that is already expired, it can request the complete history. It stores this response and can use it later on to verify any signature that was created using this certificate. The client will never need to request any status information for this certificate again. If we operate a server that stores signed documents, it can easily get the status history of certificates. Most certificates will have a simple life-cycle. They will be issued and used until they expire. The server can store this information efficiently with respect to memory consumption and performance.

3.   CONCLUSION

As this paper shows, it is much harder to verify signatures as to create them. Simple signed documents without the complete validation data cannot be automatically verified after some time, especially after the signing certificate expired. Using current revocation checking mechanisms, like CRLs and OCSP services, it may be impossible to get the status of an expired certificate. To verify a signature however, it is required to find out, if the used signing certificate was valid at the time the signature was created. Moreover, CRLs do not provide the current revocation status; even OCSP responders may not provide this information.

In principle, there are two ways to cope with this problem, if there are only OCSP or CRLs available for revocation checking. First, we can solve the problem at the time of signature creation. We can simply attach all data to the signature that is required to verify it. This includes the complete certificate chain and all corresponding CRLs or OCSP responses to verify the certificate with respect to the signing time. A big drawback of this solution is the relatively big amount of extra data that we must attach to a signature. Second, we can implement the verification software to collect all revocation information that it might need some time. This solution is not suitable for simple signature verification software; it is rather suitable for dedicated servers that store signed documents. For such servers it would be feasible to collect such big amounts of revocation information and archive them.

Nevertheless, advanced services for certificate status checking could make the verification of signatures much easier. It is not hard to design a service that provides the answer to the simple question: "Was this certificate valid at this time?" The real work for establishing such a service would be to setup a database that contains the status history for all certificates in scope. Moreover, it would be easy to design such a service in a manner that it can provide the complete history of a certificate's status.

For the future it would be desirable to have such advanced services which provide the information that the client needs to verify a certificate. While providing the required information, new services should refrain from supporting optional features which provide no real benefit. Nowadays, most of the work for revocation checking is delegated to the client, even if a server could do this work more efficiently. Today, the client has to download and archive CRLs and OCSP responses. It is really time for software developers and service providers to realize that there are serious deficiencies in current PKI (Public Key Infrastructure) technologies. Many of the current problems can be solved much better with solutions that are simpler than the current ones.

REFERENCES

[1] European Directive on Electronic Signature. The European Parliament and the Council, Brussels, December 1999, available online at
<http://europa.eu.int/ISPO/ecommerce/legal/digital.html>
[2] The Austrian Signature Law. The National Council of Austria, Vienna, August 1999, available online at <http://www.a-sit.at/>
[3] Electronic Signature Formats. Electronic Telecommunications Standards Institute, TS 101 733 V.1.3.1, France, February 2002, available online at
<http://www.etsi.org/SEC/el-sign.htm>

[4] XML Advanced Electronic Signatures (XAdES). Electronic Telecommunications Standards Institute, TS 101 903, France, February 2002, available online at <http://www.etsi.org/SEC/el-sign.htm>

[5] Housley, R., Ford, W., Polk, W., Solo, D.: Internet X.509 Public Key Infrastructure, Certificate and CRL Profile. The IETF, RFC 3280, April 2002, available online at <http://www.ietf.org/rfc/rfc3280.txt>

[6] Myers, M., Ankney, R., Malpani, A., Galperin, S., Adams, C.: X.509 Internet Public Key Infrastructure, Online Certificate Status Protocol – OCSP, RFC 2560, June 1999, available online at <http://www.ietf.org/rfc/rfc2560.txt>

[7] Adams, Carlisle, Lloyd, Steve, "Understanding the Public-Key Infrastructure", New Riders Publishing, ISBN: 157870166X

[8] Housley, R.: Cryptographic Message Syntax. The IETF, RFC 2630, June 1999, available online at <http://www.ietf.org/rfc/rfc2630.txt>

[9] XML Signature. The W3C and the IETF, Recommendation, 12 February 2002, available online at <http://www.w3.org/Signature/>

DIGITAL SIGNATURES AND ELECTRONIC DOCUMENTS: A CAUTIONARY TALE*

K. Kain
Dartmouth College
Hanover, New Hampshire, USA
kunal@Dartmouth.edu

S.W. Smith
Dartmouth College
Hanover, New Hampshire, USA
sws@cs.Dartmouth.edu

R. Asokan
Virginia Tech
Blacksburg, Virginia
rasokan@vt.edu

Abstract   Often, the main motivation for using PKI in business environments is to streamline workflow, by enabling humans to digitally sign electronic documents, instead of manually signing paper ones. However, this application fails if adversaries can construct electronic documents whose viewed contents can change in useful ways, without invalidating the digital signature. In this paper, we examine the space of such attacks, and describe how many popular electronic document formats and PKI packages permit them.

Keywords: PKI, digital signatures, e-commerce, e-government.

1.   THE PROBLEM

One of the most common uses of public-key cryptography is for *digital signatures.* If Alice performs a private-key operation on an electronic object O

---

*This work was supported in part by by the Mellon Foundation, by Internet2/AT&T, and by the U.S. Department of Justice, contract 2000-DT-CX-K001. The views and conclusions do not necessarily represent those of the sponsors.

(usually via hash and padding functions) to yield a signature $S(O)$, then those who believe in the PKI can verify that $S(O)$ came from O via Alice's public key—and thus conclude that Alice generated this signature.

Typically, O itself may consist of an object O_1 and a field I indicating intention, such as "Alice approves O_1" or "Alice witnesses that O_1 arrived by some point in time." In the process of signing, Alice takes O_1, adds I, and signs the result: indicating her approval or witness.

In the non-digital world, people must frequently take such action on *paper* documents: e.g., signing forms and expense sheets and contracts; recording when a bid or homework assignments was submitted. The last decade has seen a revolution: these paper documents have migrated into electronic settings; rather than dealing with a paper expense form, we deal with an Excel spreadsheet. This change in media allows a revolution in the ease and speed of creating and sharing documents, even between parties on opposite sides of the Internet.

The natural question arises: since we often need to sign paper documents, and a PKI would permit a nice way to sign electronic objects, can we compose the two, and indicate personal approval of a "virtual" paper document, by digitally signing the corresponding electronic object? The ability to do this composition is often a main motivating factor in deploying enterprise PKI (such as ACES [15]); businesses and products exist to provide exactly these services (we have surveyed some for this project [2, 3, 5, 9, 10]).

Typical electronic workflow tools replace a paper document with an electronic object O, that yields a virtual piece of paper when a party opens the object. When Alice signs an object O, she commits to the virtual piece of paper she sees when she views the object. By PKI, when Bob verifies $S(O)$, he concludes that Alice digitally signed O. By composition—if we accept that digital signatures imply workflow approval—Bob concludes that Alice approves the virtual paper document that O represents. Figure 1 sketches this scenario.

We decided to investigate whether this composition actually works. The particular angle that struck us as questionable was the implicit assumption that electronic objects generate semantically equivalent virtual documents each time they are opened. If $V_B(O)$ can be made to be substantially different from $V_A(O)$, then Bob's conclusion about what Alice signed will differ from what Alice thought she signed.

In Section 2, we discuss scenarios in which an adversary (perhaps Alice, or perhaps the entity who gave Alice the document) can benefit by constructing such objects. In Section 3, we discuss how popular workflow formats permit the construction of such objects. In Section 4, we discuss how many common ways of integrating PKI with workflow can still appear to validate signatures on such objects (although some packages resisted our attacks). In Section 5, we consider some countermeasures.

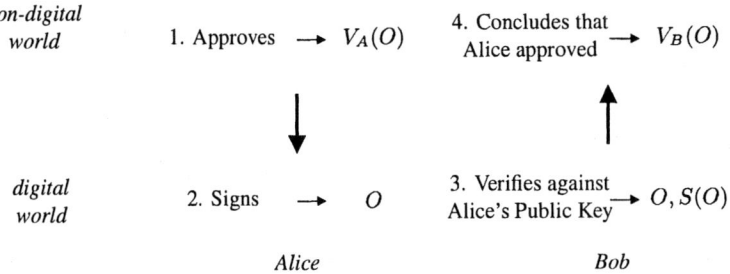

Figure 1.   Current PKI/workflow integration attempts to capture the non-digital process of signing paper, by digitally signing the corresponding electronic object. To "sign" the virtual document $V_A(O)$, Alice digitally signs the corresponding electronic object O. If Bob receives O and $S(O)$ and verifies this signature, then he concludes that Alice approved the virtual document $V_B(O)$ that he sees.

1.1   Prior Work

Our work was motivated by planned deployment of real applications that used digital signatures on electronic documents. We consequently began investigating the surprising extent to which these these document formats were malleable, and to the surprising extent to which COTS PKI packages tolerated this malleability.

However, these issues certainly have an older history. For one example, Herzberg [8] considered signing the viewing program as well as the document. For another example, we can consider Europe. The European Union's electronic signature directive [4] made electronic signatures at least as binding as paper-based signatures. Germany actually had a digital signature act (1997) even before the EU's Directive;in fact the directive was adjusted to conform to the German Digital signature Act [13]. Austria has also fully implemented the directive[1]; and concretized the malleability problem by specifying that only data formats may be used which have an "available specification" and which exclude "dynamic changes" or "invisibilities."

Pordesch [11, 12] has also considered these issues.

2.   ATTACK SCENARIOS AND APPROACHES

2.1   Motivating Scenarios

When performing vulnerability analyses, if one discovers some avenue to cause the system to engage in unexpected ways, one typically encounters the response that no one would ever do that. Consequently, we begin by pro-actively addressing this concern.

Within our university, two applications motivating campus PKI are *timestamped homework submission* and *signing of payroll and expense forms*.

- The timestamped homework application typically posits that a student Alice submits homework to an automatic system Bob that appends the current time, digitally signs the result, and forwards both on to the instructor Cathy. (One could also easily imagine a system using more advanced timestamping techniques. [7].)

  Suppose student Alice could construct an electronic document O that, whenever it was viewed, first consulted a remote file or Web location, and rendered a virtual document based on the contents of that remote file R. Alice could then completely subvert the purpose of timestamping by submitting O before the assignment deadline, but continuing to work on R up to the time the project was graded. If the assignment is one where the instructor Cathy posts sample solutions before grading, Alice could guarantee herself high scores by preparing R to follow those solutions.

- In the typical processing of expense forms, a submitter Alice sends a form to an approver Bob, who then sends it to Cathy, who actually issues checks. Suppose malicious Alice has two sets of numbers: a set S_1 with illegitimate expenses that Bob would not approve but which would result in Cathy issuing a large check, and a set S_2 with smaller numbers that Bob would approve. If Alice could construct an electronic document O that displays S_2 when viewed by Bob but S_1 when viewed by Cathy, then Cathy will receive a Bob-approved expense form indicating S_1, and Alice will a larger reimbursement than she should.

In the above two examples, Alice benefits by obtaining Bob's signature on an electronic object that, when viewed in some contexts, displays a virtual document that Bob would not have signed. Scenarios also exist where Alice could benefit by being able to retroactively change documents to which she previously committed. For example, Alice might submit a signed bid to provide (or purchase) services, and might benefit from retroactively changing the terms of that bid.

The domains of commerce, legal processes, and medical processes offer many other examples.

2.2 Taxonomy of Approaches

We now sketch a rough taxonomy of ways an adversary might construct such malleable documents.

Hidden Parameters. This avenue of attack works when the virtual document that appears when someone views an electronic object O is not completely determined by O alone, but instead depends in part on other parameters. One way to characterize attack strategies is to consider these parameters. We offer some:

- **Time.** Can the virtual document usefully change depending on the time the object is viewed?

- **Viewer Data.** Can the virtual document usefully change depending on the identity of the viewer, their machine, their operating system, or other such context data?

- **Viewer Action.** Can the virtual document usefully change depending on actions the viewer takes?

- **Remote Control.** Can the virtual document usefully change depending on the existence or contents of a file controllable by the adversary? A secondary issue here would be the proximity of this file to the viewer: a strategy that permitted the file to be an arbitrary URL would require only that the attacker control a web site, but also would require that the viewer have a good connection to the Web (in some cases, we have been able to overcome this restriction by using a 1x1-pixel image to pre-load the required document in the viewer's cache). Having the file to be closer to the viewer may require the attacker to have greater access.

Note that a Web-based remote control attack can potentially be used to mount a viewer-specific attack, if the viewer visits from a predictable host. (However, we did not try this in our tests.)

Fraudulent Content.   In our attack scenarios, the adversary devises a way to usefully change the apparent content of a signed document. The question then arises of *when* the adversary must determine this alternate content. Two natural choices suggest themselves:

- **Pre-signature.** The alternate content must be fixed at the time the signature is applied.

- **Post-signature.** The alternate content may be chosen at some point after the signature has been applied.

Nature of Change.   Another parameter is how the display of alternate content affects the "current" electronic document.

- **Static Content.** The most powerful attack strategy is one where viewing alternate content does not change the working object.

- **Dynamic Content.** A strategy that requires the working object (e.g., the Word document the viewer opened) to be modified as part of displaying alternate content can still be effective, but only when the signature is verified against the original object.

- **Dynamic Content and Signature.** It is also conceivable that an attack strategy might change signatures at the same time it changes the contents of the object, perhaps by shipping pre-established object-signature pairs.

Other Angles.   We intend the above simply as an illustrative taxonomy, not as a complete one. In particular, we also want to leave open other avenues for attack. For one example:

- **Spoofed Signature.** In standard Web spoofing [6, 16, 17], the adversary uses the richness of the user interface to create the illusion of the desired result.   When attacking signatures, the adversary might use a similar technique: the document might modify itself and invalidate its bona fide signature—but mimic the signature-verification user interface sufficiently well that the user is still convinced the signature is valid.

3.   WORFLOW FORMATS

No one is satisfied with ASCII text anymore.

In this section, we consider various popular formats for electronic documents, and the potential to realize the above attack scenarios in these formats.

We note that this is an exploratory list, not an exhaustive one. (We just wanted to see what we could find; by no means can we assert that this is complete list of all attack strategies.)

3.1   Word

3.1.1   With Macros.   Since release 6.0, Microsoft Word has permitted users to add active content to documents via *macros*. In our first attempt, we explored whether we could use Word macros to carry out a *remote control, post-signature* attack. With some trial and error, this is easily done: with the opening of the document, we associate a macro that replaces the current contents with the contents of a remote file.

```
Private Sub Document_Open()
Set doc = Documents.Open(URL of file)
Set rng2 = doc.Range
Documents("signed.doc").Activate
Set rng = ActiveDocument.Range
rng.FormattedText = rng2.FormattedText
doc.Activate
ActiveDocument.Close
```

This works, but is brittle: the viewing party must either have macro security to set to low or the attacker must be a trusted macro signer; furthermore, this is a *dynamic content* attack that changes the file contents.

3.1.2   Fields.   The security risks of Word macros are sufficiently well-known that many users leave their Macro security setting at "high"; mounting a signature attack via this vector is not particularly plausible—nor interesting. Instead, we decided to look at approaches that did not use macros.

One set of promising techniques is the *Insert Field* feature in Word. From the *Insert* menu, if one selects *Field*, one is given a rich set of fields and operators with which to build active content.

For example, we can carry out a *time* attack by using the conditional *IF* operator on the *DATE* field. In the fragment below, the author revises his testimony after the 16th. (A more complex conjunction operation would give us month and year checks as well.)

```
{ IF { DATE \@ "d" \* MERGEFORMAT } < 16
"I did not have" "I did have"}
```

Installing field code is tricky: one cannot simply "type" the code into the field box. For code such as the example, we first selected the *IF* operator from the menu, then *within* the resulting code, we inserted the *DATE* field.

The above attack has the limitation of being *pre-signature* but the advantage of being *static content*: the binary document appears to contain the entire conditional, and does not appear to change depending on the branch taken.

Fields also offer promising hooks such as *USERNAME*. However, apparently, only *DATE* and *TIME* are automatically updated upon document open; for *USERNAME* to be updated to reflect the name of the current viewer, the viewer must explicitly update fields. (although the attacker can configure the document to do this when printing).

Fields can also be updated automatically by via macros, but that reduces us to the previous case.

```
Sub UpdateAllFields()
    Dim aStory As Range
    Dim aField As Field
    For Each aStory In ActiveDocument.StoryRanges
        For Each aField In aStory.Fields
            aField.Update
        Next aField
    Next aStory
End Sub
```

(We did not examine modifying Word documents via a binary editor to see if we could cause fields to behave in an undocumented fashion.)

3.1.3   Links.   Somewhat unexpectedly, Word also permits the user to insert material from remote documents by reference. To do this, the user copies text from one Word or Excel document (we have not tried other Office formats); then, in the target document, the user selects *Edit*, then *Paste Special*, then *Paste Link*, then pastes the text in as a *link* (*unformatted text* works nicely).

This approach permits a *post-signature, remote control* attack. Unfortunately, this also appears to be a *dynamic content* attack—Word asks whether to save the document.

A surprising side-effect of the existence of this feature in Word is that a remote Web site can track each time one reads a document, and even plant cookies. (The University of Denver [14] has also noticed this "feature".)

3.2   Excel

3.2.1   Macros.   As with Word, Excel also has powerful Macro capabilities. Again, we decided not to explore this direction, because the risks are already well-known.

However, the plausibility of Excel macros as an attack vector might be greater than Word's. Many organizations use macro-laden spreadsheets as standard practice. For example, universities preparing grant proposals for the US *National Science Foundation (NSF)* are required to download Excel spreadsheets with Macros. Since these spreadsheets typically get routed throughout university administrative staff, at least one large population is primed to always hit the "accept macros" button.

3.2.2   Time and Date.   In Excel, the user can associate functional behavior with specific cells. This behavior has interesting potential for our purposes.

For example, an attacker can mount a *pre-signature, time-based* attack by selecting *Insert*, then *Function*, then building an *IF* construct using *NOW( )*. A simple example:

```
IF(DAY(NOW())<16, 2000,20000)
```

Unfortunately, this is a *dynamic content* attack, since Excel appears to notice the cell is "volatile" and asks whether the document should be saved.

3.2.3   Operating System.   We can also use functions to mount *viewer-data* attacks—although the most useful viewer data we could find was operating system (or perhaps file path name).

For example:

```
IF(INFO("osversion")<>"Windows (32-bit) NT 5.00",
"I love Linus","I love Bill")
```

Again, this attack is *pre-signature*, but also is *dynamic-content.*

3.2.4   Links.   The same techniques of Section 3.1.3 apply to Excel as well: the attacker can copy data from a Word or Excel file under his control, then *Paste Special* a link into an Excel cell. As before, this enables a *remote-control, post-signature, dynamic-content* attack.

Unfortunately, this appears to trigger a warning each time the file is opened. Options exist to disable this warning (*Tools*, then *options*, then *Edit*), but these appear to remain with the installation of Excel, rather than traveling with the file.

3.2.5   External Queries.   Excel includes features to make explicit queries to remote files. These features enable *post-signature, remote control* attacks that have *dynamic content*.

From the *Data* menu, the attacker can select *Get External Data* and then set up a query to a remote text file. The text file should be written with tab spaces between words to specify different fields in the spreadsheet. By right-clicking on the cell and selecting *Data Range Properties*, the attacker can configure the query to update on open or even regularly (in the background).

Sometimes, this technique gives a warning and an external data toolbar pop-up.

3.3   PDF

Adobe's *Portable Document Format (PDF)* is fast becoming an alternative to Word as the *de facto* standard for electronic documents.

Various tools exist to view PDF and to convert other formats into PDF; however, using the official Acrobat 5 product, one can more directly explore nuances in creating PDF documents.

With this flexibility—and with Acrobat's use of Javascript with event-driven actions—we were able to explore some interesting avenues (although we have not been able to carry out a remote-control attack here).

3.3.1   Time.   Using some Javascript functions we were able to make attacks similar to our time-based attacks on Word and Excel.

For example, one can use the *form* toolbar to create a form field, and then add Javascript code in its *calculate* field to change the value of the field according to the date.

```
var f = 9000
var g = util.printd("d", new Date())
if(g < X) f = 5000
event.value = f
```

The variable g here just gets the day out of the date value, and the X represents some day against which we want to check the value of g. In this type of form field, we can make it appear without a border—like an ordinary text. We can also make it "Read-only" using the *appearance* tab.

3.3.2   Viewer Action.   Above, we made a form that appeared to be text. We can also make a form that is invisible—but which has Javascript associated with viewer actions.

For example, when the mouse moves over the form, we can trigger Javascript to change the value of another field. We click on the *form creation* tool on the tool bar and use the mouse to create a box in the document by dragging it across the screen to the size we want the form box to be. Then , we click the *mouse enters* from the *Actions* bookmark, and select *Run Javascript* from the drag-down menu. We could then use the Javascript to give some form named "danger" another value.

```
event.value = 1
```

3.4   HTML mail

As with documents, people are not satisfied with just plain ASCII mail, and instead want mail content to incorporate feature such as colors, different fonts, and pictures. The MIME standard permits email to be formatted as HTML, and consequently many popular clients (such as Yahoo Mail, Hotmail, iname.com, and Mailcity) do this.

HTML email provides a rich breeding ground for a variety of attacks. To test these techniques, one needs to discover how to convince one's mail client to send an arbtitrary HTML file as MIME mail; in Microsoft Outlook, when sending new mail, one clicks on the *view* menu item, then selects *source edit*, then selects the *HTML* tab.

3.4.1   Date.   Using Javascript as a tool we carried out a variety of attacks in this area.

For example, we can change the content of the document with a change in date by using the `document.write()` method which allows us to write a new page using HTML tags. With many mail clients (including Netscape and Outlook), if the user attempts to look at the HTML source for this mail, they only see the final HTML—and not the Javascript with the dynamic content.

```
<HTML>
<HEAD>
<TITLE>HTML Mail</TITLE>
<SCRIPT language=JavaScript >
function InitForm()
{
var today = new Date()
var day = today.getDate()
if (day > 8)
{
  document.write("<PRE><BR>Kunal    4000<BR>Sean    8000<BR></PRE>")
}
else {
  document.write("<PRE><BR>Kunal    8000<BR>Sean    16000<BR></PRE>")
}
}
</SCRIPT>
</HEAD>
<BODY onLoad="InitForm();">
```

```
<PRE>
</PRE>
</BODY>
</HTML>
```

We can similarly do changes based on time.

3.5   Remote Control and Viewer-Specific Attacks

Another type of attack that we can carry out is include, in the HTML, references to inline images which reside on a remote system. This technique permits post-signature remote-control attacks (since we can change the images after the fact) as well as viewer-specific attacks (since our Web server can track which user is requesting the images).

We successfully carried out this attack with an image that looked like text.

```
<HTML>
<HEAD>
<TITLE>HTML Mail</TITLE>
<BODY>
<PRE>
<IMG name=thumbnail
     src="http://www.cs.dartmouth.edu/~kunal/trial1.jpg">
</PRE>
</BODY>
</HTML>
```

As noted, it is possible to get all kinds of information from already existing methods from the browser type to the operating system being used.

We can do this at the client-side by using Javascript:

```
<HTML>
<HEAD>
<TITLE>HTML Mail</TITLE>
<SCRIPT LANGUAGE="JavaScript">
function InitForm()
{
if (navigator.appName == "Microsoft Internet Explorer")
{
  document.write("<PRE><BR>Kunal    4000<BR>Sean    10000<BR></PRE>")
}
else {
  document.write("<PRE><BR>Kunal    8000<BR>Sean    16000<BR></PRE>")
}
}
</SCRIPT>
</HEAD>
<BODY onLoad="InitForm();">
<PRE>
</PRE>
```

```
</BODY>
</HTML>
```

Using Microsoft Outlook, which uses the Microsoft Internet Explorer to view HTML mail, we were able to successfully carry out this test.

4.   METHODS OF PKI/WORKFLOW INTEGRATION

We now consider some sample approaches used in the field for signing virtual documents.

4.1   External PKI

One approach is to use a PKI package that is completely oblivious to documents it is signing. For example, one might use PGP to sign and verify documents sent as email attachments. For another example, we also obtained a certificate from Digital Signature Trust [2] and used their *CertainSend* application to sign, transmit, and verify signatures on electronic documents. With these approaches, the above attack strategies on Word, Excel, and PDF work.

We also worked with S/MIME and found it vulnerable to the HTML email attacks. Experimenting with Outlook and using its built-in signature and encryption features, we were able to carry out the attacks proposed in Section 3.4 successfully.

4.2   Assured Office

Another approach is to use a PKI package that is explicitly integrated with the document software. For example, the *Assured Office* product from E-Lock Technologies [5] adds sign and verify buttons in one's Office installation, and uses a key pair resident in the user's browser to sign documents.

With this approach, all of the Word and Excel attacks we outline above still work. Even the dynamic-content attacks work, because the signature is verified at document open, before the changes have been made.

Table 1.   A summary of the attacks we tried

Types of attacks	Data Format Used
Time/Date-Based Attacks	Microsoft Word/Excel, Adobe Acrobat, HTML mail
Macro-Based Attacks	Microsoft Word
Linked file Attacks	Microsoft Word/Excel, HTML mail
Platform-Based Attacks	Microsoft Word/Excel, HTML mail
Event-Based Attacks	Adobe Acrobat

(E-Lock has recently been acquired by Lexign [9], and the Assured Office product has been renamed ProSigner.)

4.3   Acrobat Signed PDF

Adobe Acrobat has two types of signature features built in.

4.3.1   Visible Signatures.   In Adobe *visible signatures*, the document's signature is visible as an image that indicates whether the signature has been verified and whether the document was modified since. Our time-based attack of Section 3.3.1 still worked (although some versions with updates every second seemed to change the document before the signature was verified).

In theory, we could also use the techniques of Section 3.3.2 to add mouse-over action to the digital signature field—for example, to change a global value when the user verifies the signature, so interesting things can happen after the signature is verified.

The visible signature approach is also potentially vulnerable to plain old spoofing: including an image that looks just like a valid, verified signature.

4.3.2   Invisible Signatures.   In Acrobat's *invisible signatures*, the signature is applied as an invisible tag. This technique extremely well against the attacks we have proposed, as the *image* of the document can easily be matched after a change in the file content with the previous image and the changes can be easily noticed.

4.4   Utimaco

Another product we have recently discovered is the *SafeGuard Sign&Crypt for Office* from Utimaco Safeware [3], in Germany. Although we have not been able to obtain sample tools, we were pleased to notice that their online documentation indicated they were concerned about "invisible dynamic content," and the screenshots indicate they appear to sign and verify TIF images of documents. We speculate that our attack strategies would not be effective here.

4.5   Silanis

Silanis [10] of Canada makes a *Approve It* package for Word and Excel documents. In our tests, ApproveIt seems to be aware of most of our attempts to change document contents, and blocks the field contents to be updated. However, our tests also showed that ApproveIt still permits macro-based attacks.

5.   CONCLUSIONS AND COUNTERMEASURES

We consider several general approaches for potential countermeasures:

- **Inert Documents** One approach is to carefully design a document format that is inert: has no dynamic content. (This appears to be the approach taken by Utimaco.) Given the surprising richness of document formats, we would hesitate to certify any given one as inert. We also wonder whether users would accept inert documents in their workflow process.

- **Application Awareness** Another approach is to make the digital signature tools highly application aware, and refuse to verify (or perhaps even sign) documents that had malleable content. Acrobat Invisible Signatures appear to move in this direction. The fact that Excel even flags certain of our tricks as "volatile" and notices the document changes suggest an avenue to correct those issues.

  However, this approach would complicate the acceptable API that PKI tools should offer to applications.

- **Identify and Sign Parameters** Herzberg [8] once proposed signing both a document and the program that views the document. One might extend that to explicitly identify and sign all hidden parameters; however, it has been observed that this may not be a feasible solution.

- **Verification Cleanrooms** Alternatively, one might design "safe" places, free of predictable influences, where a document might be verified. For example, our remote-control Web attacks can be detected if the verifier has a slow or non-existent Web connection. (Of course, this notion of moving *verification* to a trusted safe place runs counter to the standard intuition of moving *signatures* there.)

We note that some products we tested resisted many of our attacks. and that the US appears to lag behind Europe with regard to laws and standards in this area. We would urge application deployers to carefully examine their tools.

What additional lessons do we draw from this?

First, this work offers more data points in support of standard security canards:

- The *composition* of apparently reasonable systems in not necessarily secure. What other things are dangerous to simply sign?

- As we saw before with web spoofing, the *surprising functionality and interoperability* of common desktop tools yields many opportunities for malicious behavior.

Second, by a simple exploration, we've stopped our university from deploying a fatally flawed signature/workflow integration; we offer this paper as a caution for others considering digital signatures on paperless transactions.

ACKNOWLEDGMENTS

We thank our colleagues in the Dartmouth PKI Lab for the support, and the anonymous referees for their helpful comments.

References

[1] C. Brenn. *Summary of the Austrian Law on Electronic Singatures.* http://rechten.kub. nl/simone/brenn.htm

[2] Digital Signature Trust. *CertainSend Security: A Brief Technical Overview.* http://www. trustdst.com/prod_serv/certainsend/tech_overview.html

[3] D. De Maeyer. *Interoperability at Utimaco Safeware: Digital Transaction Security.* http: //www.utimaco.de/eng/content_pdf/pkic.pdf

[4] *DIRECTIVE 1999/93/EC OF THE EUROPEAN PARLIAMENT AND OF THE COUNCIL of 13 December 1999 on a Community framework for electronic signatures.* http://europa. eu.int/ISPO/ecommerce/legal/documents/1999_93/1999_93_en.pdf

[5] E-Lock Technolpgies. *E-Lock Technologies Assured Office.* http://www.elock.com/ pdf/ao_entrust.pdf

[6] E. Felten, D. Balfanz, D. Dean, and D. Wallach. "Web Spoofing: An Internet Con Game." *20th National Information Systems Security Conference.* 1996.

[7] S. Haber and W. Stornetta. "How to Time-Stamp a Digital Document." *Journal of Cryptology.* 2:99-111. 1991.

[8] A. Herzberg. Personal communication.

[9] Lexign Incorporated. *The Lexign Suite.* http://www.lexign.com/resources/white_ papers.htm

[10] D. McKibben. *Silanis Technology: Signature Technology for E-business.* http://www. silanis.com/download/whitepapers/silanis_gartner.pdf

[11] U. Pordesch. "Der fehlende Nachweis der Präsentation signieter Daten." *DuD— Datenschutz und Datensicherheit.* 2/2000.

[12] U. Pordesch and A. Berger. "Context-Sensitive Verification of the Validity of Digital Signatures." *Multilateral Security for Global Communication* (Müller, Rannenberg, eds.). Addison-Wesley-Longman, 1999.

[13] A. Rossnagel. "Digital Signature Regulation and European Trends." http://www. emr-sb.de/news/DSregulation.PDF

[14] R.M. Smith. "Distributing Word Documents with a locating beacon." *SecuriTeam.* August 2000. http://www.securiteam.com/securitynews/5CP13002AA.html

[15] U.S. General Services Adminstration. *Access Certificates for Electronic Services.* http: //www.gsa.gov/aces/

[16] Z. Ye, S.W. Smith. "Trusted Paths for Browsers." *USENIX Security.* 2002.

[17] E. Ye, Y. Yuan, S.W. Smith. *Web Spoofing Revisited: SSL and Beyond.* Technical Report TR2002-417, Department of Computer Science, Dartmouth College. February 2002.

Index